PHYTOCHEMICALS IN
NUTRITION AND HEALTH

PHYTOCHEMICALS IN NUTRITION AND HEALTH

Edited by
Mark S. Meskin
Wayne R. Bidlack
Audra J. Davies
Stanley T. Omaye

CRC Press
Taylor & Francis Group
Boca Raton London New York

CRC Press is an imprint of the
Taylor & Francis Group, an **informa** business

Library of Congress Card Number 2001052997

Library of Congress Cataloging-in-Publication Data

Phytochemicals : their role in nutrition and health / edited by Mark S. Meskin ... [et al.].
    p. cm.
Includes bibliographical references and index.
ISBN 1-58716-083-8
    1. Phytochemicals—Physiological effect. I. Meskin, Mark S.

QP144.V44 P494 2002
613.2—dc21

2001052997

CRC Press
Taylor & Francis Group
6000 Broken Sound Parkway NW, Suite 300
Boca Raton, FL 33487-2742

© 2002 by Taylor & Francis Group, LLC
CRC Press is an imprint of Taylor & Francis Group, an Informa business

First issued in paperback 2019

No claim to original U.S. Government works

ISBN-13: 978-0-367-45500-2 (pbk)
ISBN-13: 978-1-58716-083-7 (hbk)

Visit the Taylor & Francis Web site at
http://www.taylorandfrancis.com

and the CRC Press Web site at
http://www.crcpress.com

Contents

Preface

The contents of this book evolved from papers delivered at the Third International Phytochemical Conference, "Phytochemicals: From Harvest to Health," held on the campus of California State Polytechnic University, Pomona, on November 13 and 14, 2000. Three biennial conferences have been held since 1996. These conferences have taken place during a period of time that has seen an explosion of public interest in phytochemicals and a reinvigorated and expanded growth of scientific research on plant-based bioactive chemicals The focus of these conferences has been to highlight phytochemicals that have significant potential for promoting health or preventing disease supported by solid scientific research. The conferences have also included discussions of research methodology, strategies for identifying promising chemicals, and surveys of existing research.

Data from epidemiological studies suggest that diets rich in fruits, vegetables, and whole grains are consistently associated with a decreased risk of chronic degenerative disease. The essential nutrient content alone of various plant foods cannot account for the health benefits associated with their consumption. Many researchers have correlated specific groups of phytochemicals with decreased incidence of a number of chronic degenerative disease states. The focus of current scientific investigations in the field of phytochemistry centers on the unique mechanisms of action associated with beneficial phytochemical groups and their role in the promotion of optimal health and the treatment of disease.

This third book adds new phytochemicals and foods to the list covered by the first two books, *Phytochemicals: A New Paradigm* (1998) and *Phytochemicals As Bioactive Agents* (2000) (Chapters 2, 3, 4, and 10). Topics addressed in the two previous books are also updated and expanded (Chapters 5, 7, 8, 11, and 12). Finally, several new and important topics have been added (Chapters 1, 6, and 9).

Herbal medicinal products are among the best selling nutrition-related products of our time. The public assumes these products are both safe and effective, but there is little basis for this assumption, especially in a regulatory environment in which these products are essentially unregulated. The ideal standard for determining the safety and efficacy of these plant products is the double-blind, placebo-controlled clinical trial. Such data are rarely available and when available, they are often contradictory. In Chapter 1, Ernst suggests that in the current situation the best evidence is provided by a systematic review or meta-analysis of existing data. A strong case for evidence-based herbalism is articulated and the results of systematic reviews of 11 herbal medicinal products are described. The approach described can identify herbal medicinal products that are currently safe and effective and highlight those herbs which require much more evidence prior to widespread use.

A wide range of phytochemicals has been studied as potentially therapeutic. Many of these chemicals have been identified based on chemical structure, *in vitro* activity, epidemiological evidence, and bioactivity screening programs. The fact that a chemical has interesting properties or that it demonstrates bioactivity in a screening protocol does not necessarily indicate that the chemical will be effective *in vivo* in humans. If a chemical is not bioavailable or it is heavily metabolized *in vivo*, it will be ineffective. The pharmacokinetics of phytochemicals in humans have been largely ignored. In order to understand the potential therapeutic effectiveness of a phytochemical, researchers will have to study its pharmacokinetics. Newmark and Yang suggest guidelines for approaching studies of phytochemical pharmacokinetics in Chapter 6. The illustrations provided demonstrate a wide variation in phytochemical absorption and metabolism and emphasize the need for studies of phytochemical digestion, absorption, metabolism, and excretion.

As interest heightens in the use of therapeutically beneficial phytochemicals, researchers will be looking more closely at technologies for enhancing the phytochemical content of traditional foods and for methods that will maximize plant production of phytochemicals for extraction. In Chapter 9 Still compares and contrasts the two ways plant biologists can manipulate the concentration of a phytochemical in a plant: traditional plant breeding vs. biotechnology. The methodologies, advantages, and disadvantages of both strategies are covered in detail. Bioengineering technologies have been evolving rapidly and Still's discussion clearly indicates that there is great promise in these technologies while significant challenges need to be addressed. Both novices and those already familiar with these technologies will be challenged and informed by this review.

Chapters 2, 3, 4, and 10 cover phytochemicals or foods that have not been discussed in our previous books. Echinacea-based phytomedicines are widely consumed around the world as cold and flu treatments, wound-healing agents, and immunostimulants. Many of the producers of these products and the majority of consumers do not understand or are unaware of the complex quality control issues that are inherent in the production of a phytomedicine. Arnason, Binns, and Baum in Chapter 2 point out that there are a wide number of species and varieties of Echinacea native to North America. The species that are cultivated contain a variety of phytochemicals and several medicinally important biological activities. These authors discuss such issues as botanical identity, phytochemical composition based upon species and plant part, extraction and formulation issues, and biological activity. Clearly, the quality of many Echinacea products on the market are questionable. Those interested in quality control and the production of safe and effective phytochemical-based products will benefit from this discussion.

Chapters 3, 4, and 10 are excellent reviews of phytochemicals in foods and beverages that are consumed in significant amounts by consumers and have the potential to impact human health in a meaningful way. Camire in Chapter 3 covers the fruits of the Vaccinium family which contain a variety of phytochemicals with varied health benefits. Bilberries, blueberries, and cranberries are particularly interesting because they are well liked by consumers, and it is likely that consumption of food products containing these berries could easily be increased if their health benefits can be proven. Camire discusses the anthocyanin pigments and other phytochemicals of interest, their primarily antioxidant activities, and the purported roles of these chemicals in cardiovascular disease, glycemic control and diabetes, aging and memory loss, cancer prevention, and urinary tract health. Dubick tackles a very controversial topic in Chapter 4 with a comprehensive and thoughtful survey of research on wine polyphenols and protection from atherosclerosis and ischemic heart disease. The conclusions drawn and suggestions made in this chapter should interest researchers as well as health professionals. Shahidi points out in Chapter 10 that oilseeds are an important and rich source of phytochemicals, especially phenolic and polyphenolic compounds. The occurrence and contents of oilseed phytochemicals are cataloged, health benefits and toxicology are examined, and formulation for use in foods is discussed.

Phytochemicals As Bioactive Agents (2000) included chapters on the mechanisms of chemoprevention by isothiocyanates and the effect of the soy isoflavone genistein on human breast cancer cells. Chung in Chapter 7 of this book looks at the role of isothiocyanates in cancer prevention and broadens the discussion to human clinical trials. The evidence supports the pursuit of additional research on isothiocyanates as human cancer chemopreventive agents. Kim in Chapter 8 moves the research on soy phytoestrogens into an entirely new area. The case is made that research on soy isoflavones should expand beyond cardiovascular disease, breast cancer, and prostate cancer. This chapter explores the intriguing hypothesis that soy isoflavones have neuroprotective actions in the mammalian brain. The author offers evidence from her own laboratory as well as studies from other researchers that soy isoflavones may provide protection against dementia and Alzheimer's disease. Neuroprotective mechanisms of other phytochemicals are also discussed.

Our previous two books contained a number of chapters that discussed tocopherols and carotenoids, but the primary focus was on α-tocopherol and ß-carotene. In this volume, Chapter 5 goes beyond α-tocopherol and Chapters 11 and 12 discuss research on carotenoids other than ß-carotene. In Chapter 5, Papas reviews the absorption, transport, metabolism, and biological function of non-α-tocopherols and tocotrienols and their roles in health and disease. Chapter 11 is a review of the carotenoid lycopene and its potential effects on cancer and heart disease. Lycopene is particularly interesting because it is inexpensive, widely available, and often eaten in large quantities. Burri discusses the research on lycopene and the difficulty interpreting research on a phytochemical that is already consumed in large amounts by the population. The author offers strategies for future research. In Chapter 12, Landrum, Bone, and Herrero present research on four other carotenoids: astaxanthin, ß-cryptoxanthin, lutein, and zeaxanthin. Of particular interest is the evidence of a role for lutein and zeaxanthin in protecting against age-related macular degeneration.

Our aim in this volume is to continue to support and stimulate research-based discussions of the potential role of nonessential bioactive phytochemicals in health promotion and disease prevention. It is also our goal to help focus and expand discussions of appropriate research methodologies and new technologies. We believe that the information presented in this book will be an important addition to the phytochemical and functional food literature.

MARK S. MESKIN
Editor

Editors

Mark S. Meskin, Ph.D., R.D., is associate professor, graduate coordinator, and director of the didactic program in dietetics in the department of human nutrition and food science, College of Agriculture, at California State Polytechnic University, Pomona. A registered dietitian since 1984 and a certified nutrition specialist (1995), he has been at the university since 1996.

Dr. Meskin received his bachelor of arts in psychology (magna cum laude, Phi Beta Kappa) from the University of California, Los Angeles in 1976, his master of science in food and nutritional sciences (with distinction) from California State University, Northridge in 1983, and his Ph.D. in pharmacology and nutrition from the University of Southern California, School of Medicine in 1990. A postdoctoral fellow in cancer research at the Kenneth Norris Jr. Cancer Hospital and Research Institute, Los Angeles (1990 to 1992), he subsequently served as assistant professor of cell and neurobiology and director of the nutrition education programs at the USC School of Medicine (1992 to 1996). At USC, he created, developed, directed and taught in the master's degree program in nutrition science. Dr. Meskin has also served on the faculty of the department of family environmental sciences at Cal State, Northridge, and the human nutrition program at the University of New Haven, Connecticut.

Additionally, he has several major areas of research interest including: (1) hepatic drug metabolism and the effects of nutritional factors on drug metabolism and clearance; nutrient-drug interactions; (2) the role of bioactive non-nutrients (phytochemicals/herbs/botanicals/nutritional supplements) in disease prevention and health promotion; (3) fetal pharmacology and fetal nutrition/maternal nutrition/pediatric nutrition; (4) nutrition education; (5) the development of educational programs for improving science literacy and combating health fraud.

A frequent speaker and consultant for food and pharmaceutical companies, HMOs and legal firms, Dr, Meskin is a member of many professional and scientific societies including the American Dietetic Association, the American Society for Nutritional Sciences, the American College of Nutrition, the American Council on Science and Health, the Institute of Food Technologists, and the National Council for Reliable Health Information. He is also involved in the International Life Sciences Institute, the Marilyn Magaram Center for Food Science, Nutrition, and Dietetics at Cal State, Northridge, and the Celiac Disease Foundation.

Dr. Meskin is also a member of numerous honor societies including Phi Beta Kappa, Pi Gamma Mu, Phi Kappa Phi, Omicron Nu, Omicron Delta Kappa, Phi Upsilon Omicron, and Sigma Xi. He was elected a Fellow of the American College of Nutrition in 1993 and was certified as a Charter Fellow of the American Dietetic Association in 1995. He received the Teacher of the Year Award in the College of Agriculture in 1999.

Wayne R. Bidlack, **Ph.D.**, is dean of the College of Agriculture at California State Polytechnic University, Pomona, and is professor in both the departments of animal and veterinary science and nutrition and consumer sciences.

Dr. Bidlack received his bachelor of science in dairy science and technology from Pennsylvania State University in 1966, his master of science from Iowa State University in 1968, and his Ph.D. in biochemistry from the University of California, Davis in 1972. In addition, he was a postdoctoral fellow in pharmacology at the University of Southern California School of Medicine from 1972 to 1974. Receiving an academic appointment at USC in 1974, he subsequently served as assistant dean of medical student affairs (1988 to 1991) and as professor and interim chair of pharmacology and nutrition (1992). Dr. Bidlack has also served as chairman and professor of food science and human nutrition, and as director of the center for designing foods to improve nutrition at Iowa State University, Ames (1992 to 1995).

A professional member of the Institute of Food Technologists for more than 15 years, additional affiliations include the Society of Toxicology and the International Life Sciences Institute as well as the American Institute of Nutrition (American Society of Nutritional Science), the American College of Nutrition, and the American Society of Pharmacology and Experimental Therapeutics. Dr. Bidlack has authored or contributed to more than 50 scientific publications and currently serves on the editorial board and as book editor for the *Journal of the American College of Nutrition*.

His research interests are varied but integrate the general areas of nutrition, biochemistry, pharmacology and toxicology. Specifically, he has ongoing research projects examining the hepatic mixed function oxidase enzymes, drug and toxicant metabolism and conjugation; isolation and characterization of retinoic acid UDP-glucuronosyl transferase; nutrient-nutrient interactions between vitamins and minerals, and nutrient-xenobiotic toxicities. He also maintains interest in development of value-added food products, evaluation of biologically active food components, and use of commodities for nonfood industrial uses.

In 1990, Dr. Bidlack received the Meritorious Service Award from the California Dietetic Association and the Distinguished Achievement Award from the Southern California Institute of Food Technologists. He was awarded honorary membership in the Golden Key National Honor Society in 1995 and in Gamma Sigma Delta in 1998. He also received the Bautzer Faculty University Advancement Award for California State Polytechnic University, Pomona in 1998.

Audra Davies, M.S., is manager of the department of nutrition science and services with the Access Business Group LLC, in Buena Park, CA. Her responsibilities include management of the clinical research, nutrition education, and stability programs that support the development of new products and evaluation of ingredients for the company's nutrition and wellness business line.

Prior to joining Nutrilite in 1998, Ms. Davies was director of product development for an ingredient supplier to the food and dietary supplement industries. She has a strong interest in phytochemical research and has written articles focused on natural plant chemistry and fortified and functional foods for several publications. Ms. Davies holds a bachelor of science in nutritional science (1988) and a master of science in food chemistry and nutritional biochemistry (1992) from the University of Manitoba, Winnipeg, Canada.

Ms. Davies is a professional member of Institute of Food Technologists (IFT), Canadian Institute of Food Science and Technology (CIFST), American Association of Cereal Chemists (AACC), American Nutraceutical Association (ANA), Association of Clinical Research Professionals (ACRP) and American Medical Writers Association (AMWA). She is also an active member of the Council for Responsible Nutrition (CRN).

Stanley T. Omaye, Ph.D., currently professor, department of nutrition, College of Human and Community Sciences, University of Nevada, Reno, also served as department chair and professor from 1991 to 1996.

Dr. Omaye earned a bachelor of arts in chemistry from California State University, Sacramento in 1968, a master of science in pharmacology from the University of the Pacific, Stockton, CA in 1975, and a Ph.D. in biochemistry and nutrition from the University of California, Davis, in 1975. In addition, he was a postdoctoral fellow at the California Primate Research Center from 1975 to 1976.

Dr. Omaye served as chief of the applied nutrition branch and research chemist for the biochemistry division, department of nutrition at Letterman Army Institute of Research, San Francisco, CA (1976-1980). He later served the U.S. Department of Agriculture as project leader and research chemist at the Western Human Nutrition Center, San Francisco (1983 to 1987), and project leader and research nutritionist at the Western Regional Research Center, Berkeley (1980 to 19830. More recently he was associated with Letterman as assistant chief of the toxicology division and research chemist (1987 to 1989) and research toxicologist and team leader, military trauma research division (1989 to 1991).

Dr. Omaye is a member of the American College of Nutrition, the Western Pharmacology Society, the American College of Toxicology, the Society of Toxicology, the American Institute of Nutrition, the American Society for Pharmacology and Experimental Therapeutics, and the Institute of Food Technologists. He is author or coauthor of more than 160 publications and serves on the editorial boards of *Toxicology, Society of Experimental Biology and Medicine*, and *Nutritional and Environmental Medicine*.

A certified nutrition specialist since 1993, Dr. Omaye is also certified as a Fellow in general toxicology by the Academy of Sciences (1987, 1991 and 1996), and as a Fellow by the American College of Nutrition (1993).

Contributors

John T. Arnason, Ph.D.
Department of Biology
University of Ottawa
Ottawa, Ontario

Bernard R. Baum, Ph.D.
Eastern Cereals and Oilseeds
  Research Centre
Agriculture and Agrifood Canada
Ottawa, Ontario

Shannon E. Binns
Department of Biology
University of Ottawa
Ottawa, Ontario

Richard A. Bone, Ph.D.
Department of Physics
Florida International University
Miami, Florida

Betty Jane Burri, Ph.D.
Western Human Nutrition
  Research Center
U.S. Department of Agriculture
Agricultural Research Service
Davis, California

Mary Ellen Camire, Ph.D.
Department of Food Science and
  Human Nutrition
University of Maine
Orono, Maine

Fung-Lung Chung, Ph.D.
Division of Carcinogenesis and
  Molecular Epidemiology
American Health Foundation
Valhalla, New York

Michael A. Dubick, Ph.D.
U.S. Army Institute of Surgical
  Research
Fort Sam Houston
Canyon Lake, Texas
        and
Department of Physiology
University of Texas Health
  Sciences Center
San Antonio, Texas

Edzard Ernst, M.D., Ph.D., F.R.C.P.
Department of Complementary
  Medicine
School of Postgraduate Medicine and
  Health Sciences
University of Exeter
Exeter, United Kingdom

Christian Herrero, M.S.
Department of Chemistry
Florida International University
Miami, Florida

Helen Kim, Ph.D.
Department of Pharmacology and
  Toxicology
University of Alabama at Birmingham
Birmingham, Alabama

John T. Landrum, Ph.D.
Department of Chemistry
Florida International University
Miami, Florida

Harold L. Newmark, M.S.
Department of Chemical Biology
College of Pharmacy
Rutgers, The State University of
  New Jersey
Piscataway, New Jersey

Andreas M. Papas, Ph.D.
Eastman Chemical Company
Kingsport, Tennessee

**Fereidoon Shahidi, Ph.D., F.A.C.S.,
  F.C.I.C., F.C.I.F.S.T., F.R.S.C.**
Department of Biochemistry
Memorial University of Newfoundland
St. John's, Newfoundland

David W. Still, Ph.D.
Department of Horticulture/Plant and
  Soil Science
California State Polytechnic University
Pomona, California

Chung S. Yang, Ph.D.
Department of Chemical Biology
College of Pharmacy
Rutgers, The State University of
  New Jersey
Piscataway, New Jersey

CHAPTER 1

Evidence-Based Herbalism

EDZARD ERNST

CONTENTS

INTRODUCTION

Many of today's synthetic drugs originated from the plant kingdom but, historically, medicinal herbalism went into decline when pharmacology established itself as a leading and effective branch of medical therapeutics. In much of the English-speaking world, herbalism virtually vanished from the therapeutic map of medicine during the last part of the 19th and early part of the 20th century. However, in many third world countries various forms of ethnic herbalism prevail to the present day (e.g., Ayurvedic medicine in India, Kampo medicine in Japan, and Chinese herbalism in China). In some developed countries, (e.g., Germany and France), medical herbalism continues to co-exist with modern pharmacology, albeit on an increasingly lower key.

More recently this situation has changed remarkably. The usage of medicinal herbals by the general U.S. population, for instance, has increased by a staggering 380% between 1990 and 1997 (Eisenberg et al., 1998). Medical herbalism was most commonly employed for allergies, insomnia, respiratory problems, and digestive problems. The out-of-pocket expenditure amounted in 1997 to $5.1 billion (Eisenberg et al., 1998).

Faced with this most remarkable revival of medical herbalism (also termed phytotherapy in Europe), mainstream healthcare professionals now ought to familiarize themselves with this subject. The way forward for clinicians might well be an evidence-based approach that is mainly concerned with the clinical effectiveness and safety of herbal medicinal products (HMPs).

ARE HMPs EFFECTIVE?

This seemingly simple question is not easily answered. Obviously each HMP (strictly speaking, even each type of extract) needs to be judged on its own merit. Generalizations are therefore highly problematic. Moreover, the question arises as to what type of evidence is acceptable. Traditional herbalists often seem to think that no scientific evidence is required in settling this issue, claiming that the growing and stunning acceptance of HMPs by their patients (see above) is ample proof of effectiveness. Scientists are keen to point out that traditional use is by no means proof of efficacy. In the age of evidence-based medicine, there should be no room for double standards. Proof of effectiveness can come only from (placebo) controlled, double-blind, randomized clinical trials (RCTs).

But evidence from RCTs rarely is totally uniform: one RCT shows one thing, the next another, and the next yields a different result again. Therefore, it is important not to base judgment on trials with outcomes that happen to fit preconceived ideas while discretely forgetting about data that contradict pet theories. The best way to achieve this goal is to conduct systematic reviews (and meta-analyses which are the quantitative approach to systematic reviews). Such pieces of research explain in their methods section all necessary details for rendering them reproducible. In particular, the authors have to show that selection bias was minimized and the totality of the available evidence meeting certain predefined criteria was taken into account. Traditional (narrative) reviews are today obsolete because they run a high risk of being grossly misleading, usually overoptimistic.

Researchers in the Department of Complementary Medicine, University of Exeter, have recently concentrated on conducting such systematic reviews related to defined HMPs. The results of this activity are summarized in Table 1.1. As expected, the findings vary from remedy to remedy, and often the emerging overall result is not as compelling as one would have hoped. The reasons for this are diverse. The most prevalent ones lie in the paucity of RCTs available for systematic review and in the numerous methodological flaws in trials of HMPs. Yet in some cases the evidence is sound and convincing.

The fact that systematic reviews represent the highest level of scientific evidence in terms of therapeutic effectiveness or efficacy does not mean that they are flawless or without limitations. One shortcoming is, as touched upon above, the fact that if one submits flawed RCTs to such an exercise, the result will necessarily be flawed as well. But there are other problems, e.g., publication bias. We know that negative trials tend to remain unpublished (Easterbrook et al., 1991). Thus, the published evidence can be biased toward a false positive conclusion. In our systematic reviews (Table 1.1) we invariably minimized this flaw by inviting manufacturers to contribute unpublished material. Sadly, no one can force them to do so, and we, therefore, can never be entirely sure whether all existing evidence has been included.

Recently it has been shown that the English medical literature tends to be biased toward positive results, and negative trials tend to emerge in journals published in other languages (Egger et al., 1997). It is, therefore, important to not restrict systematic reviews to the English literature. In herbal medicine this is of utmost relevance since much of the data are published in German, French, Spanish, and Chinese (for historical reasons, see above). As far as our systematic reviews are concerned, we invariably made a point of including languages other than English.

In herbal medicine there are further important problems that relate to the (lack of) standardization of extracts. If one preparation of a given herb is shown to be effective, preparation of the same herb made by a different manufacturer may not necessarily be as effective. If, subsequently, all trials of different preparations of one herb are systematically reviewed or meta-analyzed, one runs a considerable risk of generating an unreliable overall result. There is no easy way around this problem except insisting that all herbal preparations be adequately standardized. Because of this latter limitation, systematic reviews of HMPs can yield false negative (but never false positive) results.

ARE HMPs SAFE?

This question is equally complex and unanswerable through general judgments. Specific herbs will cause specific adverse effects (Ernst, 1998). The most notorious cases to illustrate the risks that can be associated with HMPs include the Belgian Chinese herbal mixture claimed to aid weight reduction which caused serious harm to over 100 individuals (van Ypersele de Strihou, 1998) and the adverse events caused by Ma huang (Ephedra) which include at least 10 fatalities (Haller and Benowitz, 2000). The reason for the Belgian cases, apparently, was a tragic misidentification of one plant in the mixture. This event emphasizes better than many words the urgent need for more stringent (quality) controls and regulation of the entire herbal sector.

Most of the users of HMPs employ these products because they (are led to) believe that HMPs are safer than conventional drug treatments. Are they wrong all the time? Obviously not. Some HMPs have been shown to be as effective and burdened with less adverse effects than synthetic competitors (e.g., Stevinson and Ernst, 1999). But no HMP is entirely free of adverse effects (Ernst et al., 2001); indeed no effective therapy will ever be. The relevant question here does not relate to adverse effects in absolute terms but to the risk/benefit relation. This relation seems tentatively positive for most of the HMPs summarized in Table 1.1.

TABLE 1.1
Systematic reviews (SR) and meta-analysis (MA) of herbal remedies*

Common (Latin) Name	Indications	Methodology	Number of Trials Included	Average Methodological Quality	Main Result (Confidence in Result)	Comment	Reference
Aloe vera (Aloe bar = badensis Mill)	Glucose-lowering, lipid-lowering, radiation injury, herpes, psoriasis, wound healing	SR	10	Poor	Some promising findings but no indication sufficiently documented (0)	Promising indications are: diabetes, hyperlipidemia, herpes, psoriasis	Br. J. Gen. Pract., 1999, 49:823–828
Complex Chinese herbal mixture	Eczema	SR	2	Good	Both trials were positive (x)	Impressive paucity of trials, no independent replication	Br. J. Pharmacol., 1999, 48:262-264
Feverfew (Tanacetum parthenium)	Migraine prevention	SR	5	Good	3 trials were positive, 2 were negative (0)	Discrepancy of results could be due to the fact that parthenolides may not be the active principle (as previously assumed)	Cephalalgia, 1998, 18:704–708
Garlic (Allium sativum)	Hypercholesterolemia	MA	13	Good (some excellent)	Overall effect is significant but of debatable clinical relevance (xx)	Methodologically best studies yield the smallest effects	Ann. Intern. Med., 2000, 133:420–429
Ginger (Zingiber officinale)	Nausea/vomiting (due to seasickness, morning sickness, chemotherapy, postoperative)	SR	6	Good	Ginger seems more effective than placebo and equally effective as synthetic drugs (xx)	Methodologically best study was negative	Br. J. Anaesthesiol., 2000, 84(3):367–371
Ginkgo (Ginkgo biloba)	Intermittent claudication	MA	8	Good to excellent	Overall positive result (xx)	Clinical relevance of effect is debatable	Circulation, 2000, 108:226–281

Common (Latin) Name	Indications	Methodology	Number of Trials Included	Average Methodological Quality	Main Result (Confidence in Result)	Comment	Reference
Ginkgo (*Ginkgo biloba*)	Tinnitus	SR	5	Poor	Only 1 trial was negative (x)	Methodologically best trial was positive	*Clin. Otolaryngol.*, 1999, 24:164–167
Ginkgo (*Ginkgo biloba*)	Vascular dementia, Alzheimer disease	SR	10	Good (some excellent)	Only 1 trial was negative	Dosage regimen varied by > 100%	*Clin. Drug Invest.*, 1999, 17:301–308
Ginseng (*Panax ginseng*)	Various	SR	25	Poor	Evidence is by and large inconclusive	Two best researched indications are physical performance and psychomotor performance/cognitive function	*Eur. J. Clin. Pharmacol.*, 1999, 55:567–575
Horse chestnut seed extract (*Aesculus hippocastanum*)	Chronic venous insufficiency	SR	8 trials vs. placebo; 5 trials vs. reference treatments	Good	Active treatment more effective than placebo and equally effective as reference treatments (xxx)	Both objective signs and subjective symptoms respond to treatment	*Arch. Dermatol.*, 1998, 134:1356–60
Peppermint oil (*Menta x piperia*)	Symptoms of irritable bowel syndrome	MA	8	Good	Positive effect of peppermint oil compared with placebo (xx)	Studies were done on enteric coated tablets: only 1 trial was negative; effect is purely symptomatic	*Am. J. Gastroenterol.*, 1997, 93:1131–5
St. Johns' wort (*Hypericum perforatum*)	Depression	SR	6 (published after MA by Linde)	Excellent	Hypericum is more effective than placebo or as effective as synthetic antidepressants (xxx)	Update of the MA by Linde; 1 trial suggested that hypericum is also effective for severe depression	*Eur. J. Neuropharmacol.*, 1999, 9:501–505
Valerian (*Valeriana officinalis*)	Insomnia	SR	2	Good	Both trials were positive	Impressive paucity of trials, no independent replication	*Sleep Med.*, 2000, 1:91-99

*0 = little confidence, x = reasonable confidence, xx = strong confidence, xxx = very strong confidence

[a] The table is confined to recent research published by the team of the author (in all cases, only double-blind RCTs have been included).

Note: For more details see Ernst, E., Pittler, M.H., Stevinson, C. and White, A. (Eds.), *Desk-top Guide to Complementary and Alternative Medicine*, Mosby, Edinburgh, 2001.

But this could be a false positive picture. Pharmacovigilance for phytomedicines is in its very early infancy (Farah et al., 2000). The fact that a given HMP has been used for millennia certainly does not prove its safety (Ernst, 1998). When trying to make sense of the safety of HMPs and comparing it to that of conventional drugs, more often than not we have to base our judgments on data which are not as reliable as we would hope. Quite simply, the existing evidence is insufficient, and it is misleading to compare the safety of synthetic drugs for which elaborate pharmacovigilance is routine with HMPs for which pharmacovigilance is but a vision for the years to come.

A particularly important problem relates to herb–drug interactions (Ernst, 2000a; Ernst, 2000b). HMPs are invariably mixtures of pharmacologically active substances, and users of HMPs are likely to also use prescription drugs. Thus, the potential for interactions is great (Ernst, 2000b), even though, due to lack of research in this area, relatively few case reports have so far emerged. Table 1.2 provides examples of herb–drug interactions that are documented in the medical literature.

CONCLUSION

HMPs can be powerful medications which may cause both benefit and harm. In order to embark on reliable risk-benefit analyses, we need to know more about their clinical efficacy and about their adverse effects. An evidence-based approach to medical herbalism seems to be the most rational way ahead. This notion is in agreement with the opinion of 88 to 90% of users of HMPs who state that research into the safety and efficacy determines their product choice (Blumenthal, 2000).

TABLE 1.2
Examples of herb-drug interactions

Medicinal Herb	Interaction	Mechanism
Garlic	Increased effect of oral anticoagulants	Garlic has antiplatelet effects
Ginkgo biloba	Increased effect of oral anticoagulants	Ginkgo has antiplatelet effects
Kava	Increased effect of other anxiolytic drugs	Synergistic action on central nervous system
St. John's wort	Increased breakdown of drugs metabolized in the liver	Hepatic enzyme inducer

Note: For more details, see Ernst, E., Possible interactions between synthetic and herbal medicinal products. Part 1: a systematic review of the indirect evidence, *Perfusion*, 2000a, 134–6 ,8-15, and Ernst, E., Interactions between synthetic and herbal medicinal products. Part 2: a systematic review of the direct evidence, *Perfusion*, 2000b, 13:60–70.

REFERENCES

Blumenthal, M., Natural marketing institute measures consumer use of herbal products, *HerbalGram*, 2000, 50:70.

Easterbrook, P.J., Berlin, J.A., Gopalan, and R., Matthews, D.R., Publication bias in clinical research, *Lancet*, 1991, 337:867–72.

Egger, M., Zellwger-Zähner, T., Schneider, M., Junker, C., Lengeler, C., and Antes, G., Language bias in randomised controlled trials published in English and German, *Lancet*, 1997, 350:326–329.

Eisenberg, D., David, R.B., Ettner, S.L., Appel, S., Wilkey, S., van Rompay, M., and Kessler, R.C., Trends in alternative medicine use in the United States, 1990–1997, *JAMA*, 1998, 280:1569-75.

Ernst, E., Harmless herbs? A review of the recent literature, *Am. J. Med.*, 1998, 104:170–8.

Ernst, E., Possible interactions between synthetic and herbal medicinal products. Part 1: a systematic review of the indirect evidence, *Perfusion*, 2000a, 13:4–6,8-15.

Ernst, E., Interactions between synthetic and herbal medicinal products. Part 2: a systematic review of the direct evidence, *Perfusion*, 2000b, 13:60–70.

Ernst, E., De Smet, P.A.G.M., Shaw, D., and Murray, V., Traditional remedies and the "test of time," *Eur. J. Clin. Pharmacol.*, 1998, 54:99–100.

Ernst, E. Pittler, M.H., Stevinson, C., and White, A. (Eds.), *Desk-Top Guide to Complementary and Alternative Medicine*, Mosby, Edinburgh, 2001.

Farah, M.H., Edwards, R., Lindquist, M., Leon, C., and Shaw, D., International monitoring of adverse health effects associated with herbal medicines, *Pharmacepidemiol and Drug Safety*, 2000, 9:105–112.

Haller, C.A. and Benowitz, N.L., Adverse cardiovascular and central nervous system events associated with dietary supplements containing ephedra alkaloids, *N. Engl. J. Med.*, 2000, 343(25):1833-1838.

Stevinson, C. and Ernst, E., Safety of Hypericum in patients with depression, *CNS Drugs*, 1999, 11:125–132.

van Ypersele de Strihou, C., Chinese herbs nephropathy or the evils of nature, *Am. J. Kidney Dis.*, 1998, 32:1-1ii.

CHAPTER 2

Phytochemical Diversity and Biological Activity in *Echinacea* Phytomedicines: Challenges to Quality Control and Germplasm Improvement

JOHN T. ARNASON,
SHANNON E. BINNS,
and BERNARD R. BAUM

CONTENTS

INTRODUCTION

Echinacea-based phytomedicines are among the most popular herbal products sold in North America and Europe (Brevoort, 1995). Diversity of products, species, and phytochemicals is a key feature of this herbal medicine. There are over 800 *Echinacea purpurea* products in Europe alone (Bauer, 1998) which are based on different plant parts and widely different extraction or formulation techniques. There is a large number of other species and varieties in the genus. The phytochemistry of the genus is diverse (several classes of phytochemicals) and redundant (many phytochemical analogues per class) and these phytochemicals provide several types of medicinally important biological activities. All of the above make the understanding and quality control and improvement of *Echinacea* products challenging.

SPECIES DIVERSITY IN *ECHINACEA*

There are nine species and four varieties of *Echinacea* native to North America according to McGregor's (1968) taxonomic treatment. The three species that are cultivated for phytomedicine production are *E. purpurea* (L.) Moench, *E. angustifolia* (D.C.), and *E. pallida* (Nutt.) Nutt., but the wild harvesting of natural populations leads to introduction of other species into commercial products. For example, we have detected *E. simulata* in commercial samples and Bauer (1998) report contamination with *Parthenium integrifolium*. Recently, we have revised the genus using numerical taxonomy based on plant morphology and phytochemistry. (Binns et al., in press). In the revised taxonomy, 10 of the 11 taxa of McGregor are retained in the revision but there is a reduction in the number of species. This revision leads to a more coherent taxonomy, and should simplify future botanical identification. A similar study of the genus using Amplified Restriction Fragment Length Polymorphism (AFLP) techniques is in progress (Baum et al., 2000). Pending publication of this revised taxonomy, the current article follows McGregor's taxonomy.

TRADITIONAL AND CLINICAL USES OF *ECHINACEA* PRODUCTS

Echinacea was one of the most important traditional medicines of the First Nations of the U.S. Great Plains and Canadian prairies. The plant, particularly the roots of *E. angustifolia*, were used mainly for treatment of sore throat, mouth sores, and septic wounds (Shemluck, 1982). Although popular in North America as an herbal medicine in the 19th century, it was abandoned in the 20th century by North Americans, and almost all laboratory and clinical evaluations up to 1990 were undertaken in Germany. Modern uses include treatment of colds and influenza, wounds, candidiasis, and lung conditions. Not surprisingly, due to the variability of preparations, clinical study results have been mixed, showing both significant and nonsignificant results of treatments. Some of the significant clinical results reported include a reduction in clinical scores and length of illness in patients with respiratory illness treated with *E. pallida* tinctures (Dorn et al., 1997), stimulation of immune responses in *Echinacea*-treated patients (Melchart et al., 1995), and positive responses in the treatment of candidiasis (Bauer, 1998).

PHYTOCHEMICAL DIVERSITY AND ASSOCIATED BIOLOGICAL ACTIVITIES

Species of the genus *Echinacea* show a high level of phytochemical redundancy with at least six different classes of secondary metabolites and multiple derivatives within any class (Table 2.1) (Bauer, 1998). The individual derivatives present vary between species and can be useful to identify the species present in manufactured products. Most biological activities have been associated with three classes of *Echinacea* phytochemicals: the caffeic acid derivatives (phenolics), the alkamides, and cell-wall derived polysaccharides.

Of the common caffeic acid derivatives (Figure 2.1), cichoric acid appears to have the greatest reported activity. It is found in appreciable amounts in *E. purpurea* (Cheminat et al., 1988). It acts as an antioxidant, it is an inhibitor of viral integrase (Robinson et al., 1996), or bacterial hyaluronidase, and it has immunostimulant activity in phagocytosis tests (Bauer, 1998). Echinacoside is used as a phytochemical marker by the medicinal plant industry and can protect collagen from free radical damage (Facino et al., 1995), but has little antimicrobial or immunomodulatory activity.

The bioactive polysaccharides are pectins and hemicelluloses in the cell wall. They are not secondary metabolites, but structural molecules that support the rigid cellulose microfibrils in a soft matrix. These molecules are readily extracted in hot water or mild base, but are not soluble in alcohol. The following polysaccharides have been identified from *E. purpurea*: an 80kDa xyloglucan, a 45kDa arabinotrhamnogalactan, and a 35kDa 4-o-methyl-glucoronoarabinoxylan (Bauer, 1998). Additional polysaccharides and glycoproteins have been characterized from *Echinacea* cell cultures. Polysaccharides from *Echinacea* have potent immunostimulant activity, i.e., macrophage activation and cytokine production (IL-1, IL-6, IL-10 TNF-alpha) (Rininger et al., 2000; Wagner et al., 1988). Their activity is comparable to other polysaccharides from immunostimulant plants, such as ginseng, but the

TABLE 2.1
Phytochemical diversity and redundancy in *Echinacea*

Class of compound	Approximate number of derivatives
Alkamides	>20
Caffeic Acid Derivatives	> 5
Flavonoids	> 5
Alkaloids	2
Polyacetylenes	>10
Polysaccharides	> 5
Essential Oil Volatiles	>15

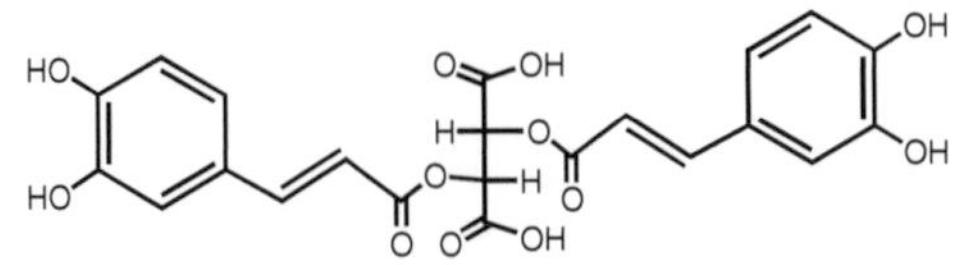

CICHORIC ACID

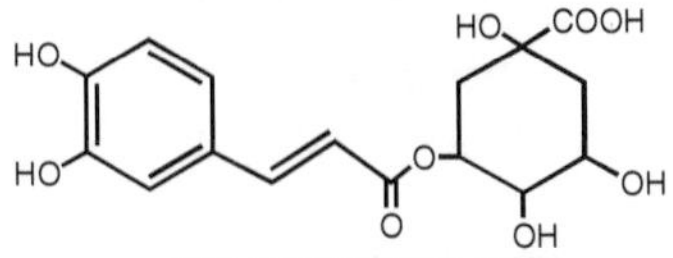

CHLOROGENIC ACID

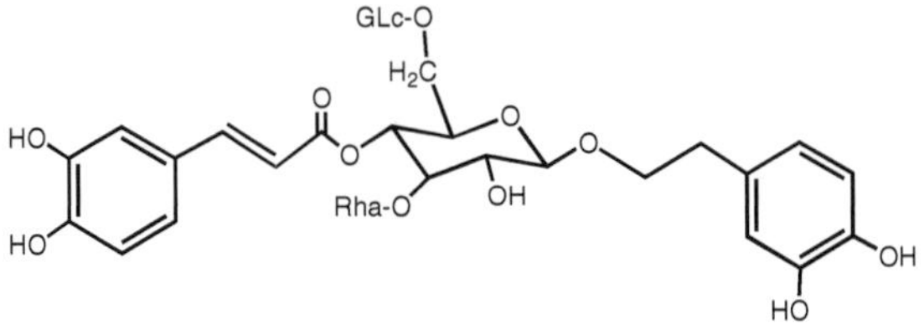

ECHINACOSIDE

Figure 2.1 Representative caffeic acid derivatives of *Echinacea.*

polysaccharide composition is different. Recent work shows that digestion (simulated) is required to express their activity (Rininger et al., 2000). The pharmcokinetics of these materials remain poorly understood.

The alkamides (isobutylamides) (Figure 2.2) and related polyacetylenes are fatty-acid derived molecules with unsaturated double and triple bond systems and are distributed in roots and flowers. *E. angustifolia* roots are an especially rich source of these compounds. The isobutylamides are the taste-tingling components of *Echinacea* that have analgesic properties and provide relief for sore throat symptoms. The isolated alkamides are also known to be mammalian lipoxygenase (LOX) and cyclo-oxygenase (COX) inhibitors (Muller-Jakic et al., 1994) which confer an anti-inflammatory activity. New research from our group has shown that lipophilic extracts of *Echinacea* containing isobutylamides and polyacetylenes are toxic to a variety of pathogenic fungi, including multidrug-resistant *Candida* spp. (Binns et al., 2000). This activity is enhanced by light and is relevant to traditional uses for treatments of sores and wounds and modern uses of *Echinacea* as a topical agent. Further studies by J. Hudson (unpublished) have shown that *Echinacea* extracts also inhibit *Herpes simplex* (cold sore virus).

Consideration of the components mentioned above is important in predicting the activity of different formulations of *Echinacea.* Teas or the expressed juice of *E. purpurea* tops used in many European products are rich in polysaccharides and phenolics, and these products in good condition should have immunostimulant and

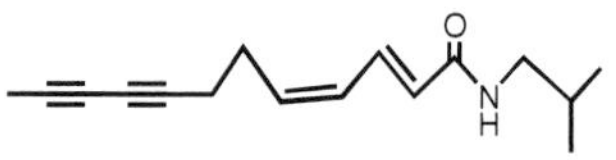

PID 7

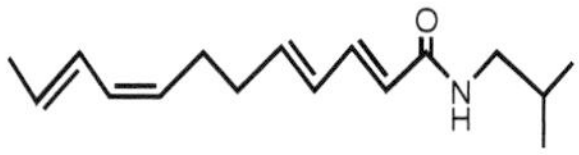

PID 8

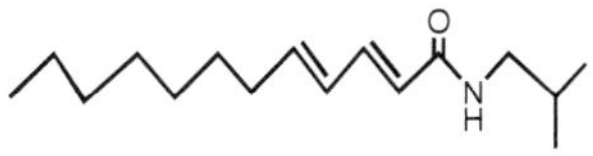

PID 11

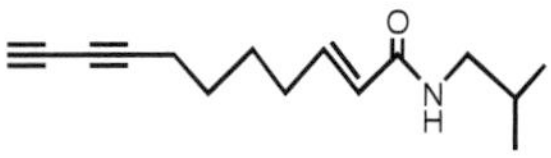

PID 12

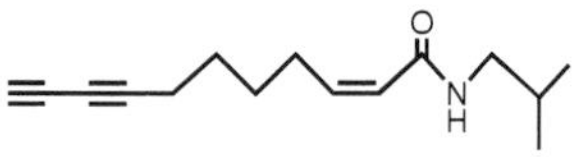

PID 13

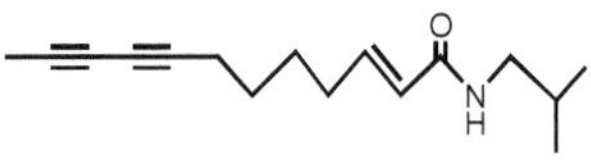

PID 14

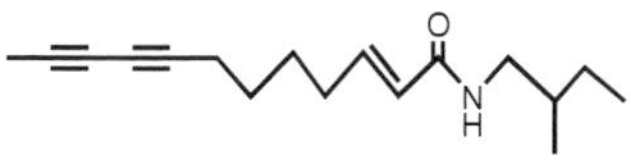

PID 17

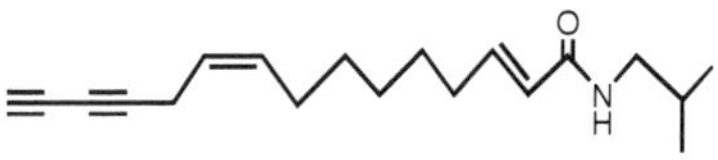

PID 18

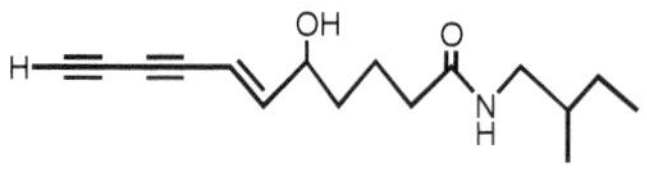

PID 20

Figure 2.2 Representative isobutylamides of *Echinacea.*

antioxidant properties. The alcoholic extracts popular in many preparations, on the other hand, have higher levels of alkamides and are expected to have analgesic, anti-inflammatory, antioxidant, and antimicrobial properties.

QUALITY CONTROL

As mentioned previously, quality control is a serious problem in *Echinacea* phytomedicines with multiple actives, species, and formulations. Early industry methods based on the determination of total phenolics are unsuitable for quality control because of the broad distribution of phenolics in all terrestrial plants. For species identification and quality assurance, high performance liquid chromatography (HPLC) determination of specific phytochemicals is the most appropriate technique (Bergeron et al., 2000; Perry et al., 1997). Extraction of most phenolic and alkamide components can be conveniently achieved in 70% ethanol using ultrasound and analyzed by gradient diode array HPLC. No effective quantitative methods are yet available for the rapid analysis of polysaccharides, but HPLC techniques and immunochemical analysis are under development.

For species identification of root materials, cichoric acid is often used in the herbal industry as a marker for *E. purpurea* and echinacoside for *E. angustifolia* and *E. pallida*. However, we have found these markers to be present in a number of wild species and varieties and are consequently not the best choices for species identification. On the other hand, the numerous alkamides found in the roots provide a phytochemical profile that is unique for each commercial species (Bauer and Remiger, 1989) (Table 2.2), and are different from wild species.

Quantitative estimates of the phytochemical markers assessed by HPLC showed a large amount of variation in commercial samples of different origin submitted to our laboratory for analysis (Bergeron et al., 2000). For example, the level of alkamides or cichoric acid showed a 10-fold variation in raw plant material. This variation can be traced to a number of factors, including genetic variation in the crop, effect of growing conditions, and loss of active principles in postharvest processing.

IMPROVEMENT OF *ECHINACEA* QUALITY

Recent studies by Livesy et al. (2000) and Bauer (1997) show that *Echinacea* markers are unstable in juice, during harvesting and drying, and in storage. However, losses can be minimized by rapid drying during harvest, protection from oxidation in storage, and use of cooler storage temperatures.

Improvement in phytochemical marker consistency and amount can be achieved in a number of ways. For example, phytochemical constituents are well known to be influenced by soil nutrients, and we have recently found (Table 2.3) that the alkamides are inducible by the plant hormone methyl jasmonate (Binns et al., 2001). In collaboration with Trout Lake Farm LLC, we found that flowering stage is also an

TABLE 2.2
Alkamides useful in species identification

	E. purpurea	*E. angustifolia*	*E. pallida*
2,4 diene isobuylamides	Many	Few	
Mono-ene isobutylamides	Few	Many	
Ketoalkene and Ketoalkyne			Many

TABLE 2.3
Alkamides and ketoalkene/ynes in *E. pallida* roots (58 days) that showed a statistically significant increase by methyl jasmonate treatment

Compound induced	% increase	P
tetradeca-8Z-ene-11,13-diyn-2-one*	151	0.016
dodeca-2E,4E,8Z,10E/Z-tetraenoic AIBA	63	0.046
trideca-2E,7Z-dien-10,12-diynoic AIBA	227	0.035
undeca-2E,4Z-dien-8,10-diynoic AIBA	134	0.021
undeca-2Z,4E-dien-8,10-diynoic AIBA	156	0.007
8-hydroxytetradeca-9E-ene-11,13-diyn-2-one*	353	0.001

Note: Significantly different means were determined at P=0.05 by a two-tailed t-test.

important determinant of cichoric acid concentration in aboveground harvest (Letchamo et al., 1999). Cichoric acid is most concentrated in immature (preanthesis) flowers of *E. purpurea* and declines after anthesis. Alkamide content of flowers has an opposite trend (Figure 2.3). Genetics is another important component. Even in germplasm that has been in cultivation for many years, there is still considerable phytochemical variation between individual plants. Cloned plants derived by division of the roots of individual plants in cultivated populations are very uniform and are one method for selecting and producing high-performance cultivars from exceptional plants. Development of tissue culture methods, currently under investigation in our laboratory, is expected to allow much faster and large-scale production of uniform elite plants.

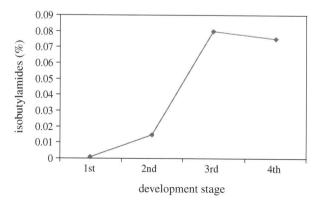

Figure 2.3 Changes in isobutylamides with flower stage. Stages 1 and 2 are preanthesis and stages 3 and 4 are postanthesis.

Wild populations of *Echinacea* provide further scope for genetic improvement. Our greenhouse investigations of *E. angustifolia* populations grown from seed collected at nine locations on the Great Plains showed a latitudinal variation in the amounts of several phytochemical markers (Binns unpublished). Echinacoside, for example, increases significantly with latitude, while several alkamides decrease as latitude increases.

NEW TOOLS FOR QUALITY CONTROL

Germplasm improvement requires assessment and selection from a large number of individuals. Recently we have developed molecular-methods based AFLP (Amplified Restriction Fragment Length Polymorphism), which reliably predict the HPLC-determined cichoric acid content across a wide selection of *Echinacea* germplasm (Baum et al., 2001). These methods are useful in the development of genetic-marker assisted selection of crop germplasm, which is expected to be much faster than traditional phytochemical analysis.

Assessment of the quality of raw materials and finished products can be enhanced with modern bioassays. Protein-based bioassays such as LOX enzyme inhibition assays by *Echinacea* or cytokine stimulation in macrophages by *Echinacea* have been considered recently for quality control (Rininger et al., 2000). We have also assessed the potential for drug interaction with different *Echinacea* herbal products in assays with cloned human CYP3A4 enzymes (Budzinski et al., 2000). New DNA array and quantitative PCR techniques now allow scientists to rapidly assess the induction of genes for cytokine production by measuring m-RNA levels in cultured cells treated with Echinacea (Orndorff, 2000). While these bioassays cannot completely replace constituent analysis, they are powerful new additions to the toolkit for assessment and improvement of herbal products.

REFERENCES

Bauer, R. (1997). Standardization of *Echinacea purpurea* expressed juice with reference to cichoric acid and alkamides (in German), *Z. Phytother.*, 18:270–276. English version in press in *Herbs, Spices, and Medicinal Plants*.

Bauer, R. (1998). *Echinacea*, Biological effects and active principles, in ACS Symposium Series, 691, 140–157.

Bauer, R. and Remiger, P. (1989). TLC and HPLC analysis of alkamides in *Echinacea* drugs, *Planta Med.*, 55, 367–371.

Baum, B.R., Mechanda, S., Livesy, J., Binns, S.E., and Arnason, J.T. (2001). Predicting quantitative compounds in single *Echinacea* plants or clones from their DNA fingerprints, *Phytochemistry*, in press.

Baum, B.R., Mechanda, S., Binns, S.E., and Arnason, J.T. (2000). *Echinacea* germplasm enhancement project, DNA fingerprinting and intellectual property enhancement, Proc. American Herbs Prod. Int. *Echinacea* Symposium.

Bergeron, C., Livesey, J., Arnason, J., Awang, D.V.C., Baum, B., and Arnason, J.T. (2000). A quantitative HPLC method for *Echinacea* analysis, *Phytochemical Anal.*, 11: 207–215.

Binns, S.E., Purgina, B., Smith, M.L., Johnson, L., and Arnason, J.T. (2000). Light-mediated antifungal activity of *Echinacea* extracts, *Planta Medica*, 66:1–4.

Binns, S.E., Inprajah, I., Baum, B.R., and Arnason, J.T. (2001). Methyl jasmonate increases reported alkamides and ketoalkynes in *Echinacea pallida, Phytochemistry,* in press.

Brevoort, P. (1995). The U.S. botanical market — an overview. *HerbalGram,* 36, 49–57.

Budzinski, J., Foster, B., Vandenhoek, S., and Arnason, J.T. (2000). *In vitro* evaluation of human cytochrome P450 3A4 inhibition by herbal tinctures, *Phytomedicine,* 274: 273–282.

Cheminat, A., Zawatzky, R., Becker, H., and Brouillard, R. (1988). Caffeoyl conjugates from *Echinacea* species: structures and biological activity, *Phytochemistry,* 27, 2787–2794.

Dorn, M., Knick, E., and Lewith, G. (1997). Placebo-controlled, double-blind study of *Echinacea pallida radix* in upper respiratory tract infections, *Complementary Therapies in Medicine*, 3:40–42.

Facino, R.M., Carini, M., Aldini, G., Saibene, L., Pietta, O., and Mauri, P. (1995). Echinacoside and caffeoyl conjugate protect collagen from free radical induced degradation. *Planta Med.*, 61, 510–514.

Letchamo, W., Livesy, J., Arnason, J.T., and Krutilina, V.S. 1999. Cichoric acid and isobutylamide content in *Echinacea* as influenced by flower development, Proc. Symp. New Crops and New Uses Conference, Arizona.

Livesy, J., Awang, D.V.C., Arnason, J.T., Letchamo, W., Barrett, M., and Pennyroyal, G. (2000). Effect of temperature on stability of marker constitutents in *Echinacea purpurea* root formulations. *Phytomedicine*, 6:347–9.

McGregor, D. (1968). The taxonomy of the genus *Echinacea*, The University of Kansas Science Bulletin, 68:113–142.

Melchart, D., Linde, K., Worku, F., Sakady, L., Holzmann, M., Jurik K., and Wagner, H. (1995). Results of five studies on the immunomodulatory activity of *Echinacea, J. Alternative and Complementary Med.*, 1:145–160.

Muller-Jakic, B., Breu, W., Pröbstle, A., Redl, K., Greger, H., and Bauer, R. (1994). *In vitro* inhibition of cyclooxygenase and 5-lipoxygenase by alkamides from *Echinacea* and *Achillea* species, *Planta Med.*, 60, 37–40.

Orndorff, S. (2000). Discovery of novel botanical ingredients using high throughput purification and quantitative gene expression technologies, *Proc. Am. Soc. Pharmacognosy*, Seattle.

Perry, N.B., van Klink, J.W., Burgess, E.J., and Parmenter, G.A. (1997). Alkamide levels in *Echinacea purpurea*: A rapid analytical method revealing differences among roots, rhizomes, stems, leaves, and flowers, *Planta Med.*, 63, 58–62.

Robinson, W.E., Reinecke, M.G., Abdel-Malek, S., Jia, Q., and Chow, S.A. (1996). Inhibitors of HIV-1 replication that inhibit integrase, *Proc. Natl. Acad. Sci.*, 93, 6326–6331.

Rininger, J.A., Kickner, S., Chigurupati, P., McLean, A., and Franck, Z., (2000). Immunopharmacological activity of *Echinacea* preparations following simulated digestion on murine macrophages and human peripheral blood mononuclear cells, *J. Leuk. Biol.*, 68, 503–510.

Shemluck, M. (1982). Medicinal and other uses of compositae by Indians in the U.S. and Canada, *J. Ethnopharmacology*, 5:303–358.

Wagner, H., Stupner, H., Schafer, W., and Zenk, M. (1988). Immunologically active polysaccharides of *Echinacea purpurea* cell cultures, *Phytochemistry*, 27:119–126.

CHAPTER 3

Phytochemicals in the *Vaccinium* Family: Bilberries, Blueberries, and Cranberries

MARY ELLEN CAMIRE

CONTENTS

INTRODUCTION

Vaccinium fruits are rich in antioxidant activity. Purported health benefits of these berries include maintenance of normal vascular health and vision, prevention or reduced severity of cardiovascular disease, diabetes and cancer, and antimicrobial action. Although many species are consumed, only a few have achieved commercial significance. These berries have an important place in the botanical history of many cultures. Increased production has led to ample supplies and consequently, commodity groups have begun to sponsor research on the health benefits of these berries in order to tap the growing functional foods and nutraceutical markets.

SPECIES

Vaccinium species in the Chinese Materia Medica include the Asiatic bilberry (*V. bracteatum* Thunb.), South China blueberry (*V. dunalianum* var. *urophyllum* Rehd. and Wils.), and bilberry or lesser bilberry (*V. fragile* Franch.). Other species have a place in traditional medicine in Turkey and in northern climates. The species discussed in this chapter will focus on those harvested for commercial use in North America and parts of Europe.

The bilberry (*Vaccinium myrtillus* L.) is native to northern Europe and is also found in parts of North America and Asia. The fruit of this low-growing shrub has an intense dark blue color with pigmented flesh. Bilberries grown in Europe have found application in the pharmaceutical and dietary supplement markets. The commercial drug Difrarel® contains 100 mg of bilberry anthocyanins plus 5 mg of β-carotene, and is prescribed for circulatory ailments.

Blueberries are native to North America and commonly found as either the lowbush or highbush variety. In both varieties, the blue pigments are confined to the berry skin and berries may be harvested mechanically or by hand. The wild or lowbush blueberry (*V. angustifolium* Ait.) is grown primarily in Maine and eastern Canada. Highbush blueberries (*V. corymbosum* L.) are commercially grown in large quantities in New Jersey and Michigan (Moore, 1994), as well as in other states that meet the temperature requirements of 800 to 1060 h below 7.2°C without winter temperatures below -24°C (Eck, 1998). In Canada, British Columbia is the largest producer of cultivated highbush blueberries (Villata, 1998) and other major growing regions include Australia, Chile, Germany, and New Zealand. The crop is also being developed in regions of Europe and China. *V. ashei* Reade, the rabbiteye blueberry, is produced in the southern U.S., with Georgia, Florida, Arkansas, and Texas being the major growing states.

Several cranberry species are grown throughout the world for commercial application. The American cranberry (*V. macrocarpon* Ait.) is produced commercially in Massachusetts, New Jersey, Wisconsin, Oregon, Maine, and in other northern states and Canadian provinces. The European or small cranberry (*V. oxycoccus* L.) and the lingonberry (*V. vitis-idaea* L.) may also be found in North America despite their predominance in northern Europe and the former Soviet republics.

MARKETING TRENDS

As international notoriety grows, horticultural production is spurred. Anticipation of new markets for *Vaccinium*-derived nutraceuticals can lead to supply problems. For example, in the 1990s cranberry production increased in the U.S. and Canada, but demand failed to rise. In 2000, farmers received only ten cents per pound, compared with $0.80 in earlier years. New plantings of blueberries and bilberries are expected to have a similar effect on prices for those berries. Generally speaking, specified health claims for phytochemical activity of berries do not exist. "Dr. Blueberry" appears in Japanese products containing blueberries for health benefits, but as yet there is no Foods for Specified Health Use (FOSHU) claim approved for blueberry products.

PHYTOCHEMICALS

The most apparent phytochemicals in these fruits are the anthocyanin pigments. Most anthocyanins are colorless at pH 4–6, and anthocyanin activity is reduced by heat, oxygen, ascorbic acid, light, and polyphenoloxidase. Anthocyanins in *Vaccinium* species are discussed in detail by Mazza and Miniati (1993). Other phenolics and flavonoids are present in varying amounts. Chlorogenic acid is the predominant phenolic compound in blueberries (Kader et al., 1996); benzoic acid is the primary phenolic in cranberry juice (Chen et al., 2001), and quercetin is found in most species as well. Berry flavonoid and phenolic composition differs greatly from that in leaves. Geographic and cultivar variations are significant in blueberries, while geographic differences may be less significant for cranberries.

ANTIOXIDANT ACTIVITY

IN VITRO STUDIES

Numerous studies have evaluated the *in vitro* antioxidant activity of *Vaccinium* constituents. Chlorogenic and caffeic acids inhibited linoleic acid oxidation (Ohnishi et al., 1994). Crude acidified ethanol extracts from four *V. corymbosum* cultivars were studied for their antioxidant activity, but this activity was not directly related to anthocyanin or polyphenol content of the extracts (Costantino et al., 1992). Cv. Collins was the most effective inhibitor of both NBT reduction by superoxide anion and xanthine oxidase, and it had the highest concentration of both anthocyanins and polyphenols.

Velioglu and coworkers (1998) compared the antioxidant activity of methanol extracts from 28 food and medicinal plants using a β-carotene bleaching assay. *V. angustifolium* cv. Fandy from Nova Scotia had greater antioxidant activity than other products studied, with the exception of 50 mg/L α-tocopherol, 200 mg/L BHT, and a few other botanical products such as sea buckthorn and buckwheat hulls.

The oxygen radical absorbance capacity (ORAC) test is an assay which can assess hydrophilic antioxidants for their ability to reduce free radicals. Blueberry and bilberry ORAC values were correlated with anthocyanins ($r = 0.77$) and total phenols ($r = 0.92$) (Cao et al., 1998), with variation among species and cultivars. The researchers concluded that ascorbic acid played only a minor role in ORAC activity.

ORAC activity, anthocyanins, and total phenolics increased with maturity in the two rabbiteye blueberry cultivars, Tifblue and Brightwell. Highbush blueberry cv. Jersey was grown in three states, but location had no effect on composition or ORAC activity. Based on these findings, the researchers suggested that daily consumption of 1/2 cup (72.5 g) of blueberries would raise ORAC intake by 1–3.2 mmol, based on Trolox equivalency. Fourteen anthocyanins tested for ORAC had greater antioxidant activity than Trolox (Wang et al., 1997). Cyanidin-3-glucoside and cyanidin-3-rhamnoglucoside had the highest ORAC values.

Oxidation of human low density lipoprotein (LDL) oxidation is involved with development of atherosclerotic plaques in arteries. *Vaccinium* phenolic compounds vary in ability to retard LDL oxidation (Meyer et al., 1998); however, additive effects were lower than expected. In the aqueous-based ABTS+ system, the antioxidant activity of compounds was reversed: quercetin > cyanidin > catechin (Rice-Evans et al., 1995).

Berry antioxidant activity in different systems was compared by Heinonen and coworkers (1998). Acetone extracts of highbush blueberries (cv. Jersey) were less effective in preventing LDL oxidation than were extracts from blackberries, red raspberries, and sweet cherries. Hydroxycinnamates in blueberries were better at inhibiting oxidation in lecithin liposomes.

When 10 μM copper was used to initiate human LDL oxidation, anthocyanins were protective in the following order: malvidin > delphinidin > cyanidin > pelargonin; at higher copper concentration (80 μM) delphinidin was most effective (Satué-Gracia et al., 1997). Malvidin was the best antioxidant in a liposome model since the others were pro-oxidants at the 3 μM level.

Aqueous extracts of *V. myrtillus* at low levels (15–20 μg/mL) retarded human LDL oxidation *in vitro* (Laplaud et al., 1997). The lag phase for production of conjugated dienes was significantly delayed by the bilberry extract, and levels of peroxides and thiobarbituric acid reactive substances (TBARS) were lower for up to seven hours after induction with copper. Further research is needed to evaluate dose levels and long-term consumption effects.

Although cyanidin is not one of the predominant anthocyanins in bilberries and blueberries, it has been studied the most. Cyanidin and cyanidin-3-O-β-D-glucoside (C3G) were effective antioxidants in several systems: linoleic acid autoxidation, lecithin liposome, rabbit erythrocyte membrane, and rat liver microsomes (Tsuda et al., 1994). The aglycone was stronger in the liposome and rabbit cell systems. Reaction of C3G with a free radical generator (2,2'-azo-bis-(2,4)-dimethylvaleronitrile) produced another free radical scavenger, protocatechuic acid, and 4,6-dihydroxy-2-O-β-D-glucosyl-3-oxo-2,3-dihydrobenzofuran (Tsuda et al., 1996).

IN VIVO STUDIES

In another study, rats were fed C3G (2 g/kg diet) for 14 d (Tsuda et al., 1998). There was no difference between the control animals and the experimental group for weight gain, food intake, liver weight, and serum triglycerides, phospholipids, and antioxidants (α-tocopherol, ascorbic acid, reduced glutathione, and uric acid). Free and total cholesterol was lower in the C3G-fed rats. Serum was subjected to *ex vivo* oxidation by two methods: 2,2'-azobis (2-amidinopropane) hydrochloride (AAPH) and $CuSO_4$. Initially, the serum TBARS from the C3G rats were slightly but significantly lower than that of the control group, but the differences became more pronounced with time in both reaction systems. C3G did not appear to have an antioxidant effect in rat liver.

A small study of healthy Danish women ($N = 9$) found no difference in antioxidant status after drinking 500 mL of blueberry juice, but increased antioxidant activity was maximal 1–2 h after drinking an equal amount of cranberry juice (Pedersen et al., 2000). In a Finnish study, 60-year old men consumed 100 g of frozen berries (bilberries, lingonberries, and black currants), 600 mg antioxidant vitamins, or a calcium placebo for eight weeks (Marniemi et al., 2000). Although decreases in LDL oxidation and increases in antioxidant activity were not significant at the 5% level, a short-term (5-hr) study found a significant increase ($p = 0.039$) in antioxidant activity following berry consumption.

CARDIOVASCULAR DISEASE

ARTERIAL RELAXATION

Cardiovascular disease, as well as the many diseases involved with vascular health, can be expected to increase in the U.S. population as the proportion of older adults grows. Schramm and German (1998) have summarized potential benefits of flavonoids in preventing vascular disease, including: prevention of oxidation and Maillard reactions, inhibition of leukocyte adhesion, antimicrobial activity that prevents inflammation, and estrogenic effects. Figure 3.1 illustrates potential cardiovascular benefits of *Vaccinium* phytochemicals.

Hypertension, atherosclerosis, and diabetes decrease arterial flexibility, contributing to poor blood flow and plaque formation. Rat aortas exposed to anthocyanin-enriched blueberry extract *in vitro* exhibited relaxation caused by endothelium-generated nitric oxide (Andriambeloson et al., 1996). Delphinidin induced a maximal relaxation of 89%, which was comparable to red wine polyphenols

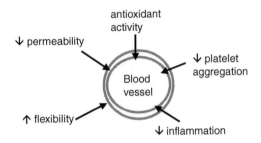

Figure 3.1 Potential cardiovascular benefits of *Vaccinium* phytochemicals.

(Andriambeloson et al., 1998). Fitzpatrick and coworkers (1995) reported that malvidin did not produce a vasorelaxation response. Since neither malvidin nor cyanidin were effective, vasorelaxation effects may be caused by as yet unknown anthocyanin features.

A commercial *V. myrtillus* extract (Myrtocyan®) enhanced relaxation of calf aortas in an *in vitro* system that had been exposed to adrenalin (Bettini et al., 1985). The dose response was nearly linear as the extract concentration was increased from 25–100 µg/mL (Figure 3.2), and the researchers proposed that the relaxation of the blood vessels was due to inhibition of catechol-O-methyltransferases (COMT). Anthocyanins and ascorbic acid inhibited contractile responses of calf aortas in the presence of histamine or angiotensin II (Bettini et al., 1987). This effect was not observed when indomethacin or lysine acetylsalicylate were added, suggesting that anthocyanin-rich products may have little benefit for persons already taking medications to inhibit prostaglandin synthesis enzymes.

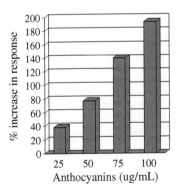

Figure 3.2 Bilberry anthocyanins potentiated relaxation effect of 0.2 µg/mL adrenalin in calf coronary arteries. (Adapted from Bettini et al. 1985. *Fitoterapia*, 56(2):67–72. With permission.)

CAPILLARY PERMEABILITY

Mian and coworkers (1977) proposed that anthocyanins protected blood vessels by stabilizing membrane phospholipids and by increasing production of the acid mucopolysaccharides of the connective ground substance. Anthocyanin extracts inhibited porcine elastase *in vitro* (Jonadet et al., 1983). Vascular protection in rats was measured by retention of Evans blue dye in serum, and appeared to be dose-dependent. Grape (*Vitis vinifera*) anthocyanins were better at protecting blood vessels than *V. myrtillus* extracts at the lowest dose, 50 mg/kg/I.P. Extracts from both fruits were significantly different from the control treatment at doses of 100 and 200 mg/kg/I.P.

Capillary strength is another important element in cardiovascular health. Easily damaged or porous capillaries contribute to electrolyte imbalances, and lead to edema and other dysfunctions. The commercial *V. myrtillus* preparation Difrarel® containing 20 mg of anthocyanins was tolerated well by patients with chronic illnesses affecting blood vessels, even though patient improvement was mixed (Amouretti, 1972). Coget and coworkers (1968) described the progress of 27 patients treated with Difrarel 20®, and concluded that the drug was an effective vascular protective agent. Exposure to radiation, either therapeutic or accidental, promotes weakened capillaries. Interest in protective effects of *V. myrtillus* against radiation damage has been high in Eastern Europe.

Oral doses of *V. myrtillus* extract were slightly less effective in increasing capillary resistance to permeability than were intraperitoneal administrations to rabbits and rats (Lietti et al., 1976a). However, hamsters given oral doses (10 mg/10 g body weight) of a commercial product containing 36% bilberry anthocyanosides for 2 or 4 weeks exhibited better capillary perfusion and fewer sticking leukocytes in the capillaries that had been clamped to induce ischemia (Bertuglia et al., 1995).

A commercial *V. myrtillus* extract, known as Myrtocyan®, with 25% anthocyanidins, was effective in reducing capillary permeability and increasing capillary resistance (Cristoni and Magistretti, 1987). The extract was administered intraperitoneally to rats starved for 16 hr, then Evans blue dye was given intravenously, followed by an intradermal dose of histamine. Rats given varying amounts of Myrtocyan had reductions of 32–54% in permeability.

PLATELET EFFECTS

A related cardiovascular issue involves platelet aggregation where excessive aggregation has been known to lead to blood vessel blockage. In a small human study, consumption of cranberry juice (4 glasses per day for 4 days) inhibited aggregation *ex vivo* (Wilson and Marley, 2001). A solution of 30 mg/ml bilberry anthocyanins reduced platelet aggregation *in vitro*, comparing favorably with aspirin and other drugs (Zaragozá et al., 1985).

Elevated serum homocysteine is yet another risk factor for cardiovascular disease. Interestingly, in a study where cranberry juice was used as a vehicle for nitroglycerine to study endothelial dysfunction, the cranberry juice placebo had no effect on flow-mediated vasodilation and serum asymmetrical dimethylarginine (Boger et al., 2001). Despite this finding, *Vaccinium* phytochemicals may play a role in these aspects of vascular health.

GLYCEMIC CONTROL AND DIABETES

Diabetes is a leading cause of death, and the incidence of Type 2 diabetes is growing in most developed nations. The leaves and fruit of *V. myrtillus* are used in Europe to treat many conditions resulting from diabetes. High levels of serum glucose trigger many adverse physiological events. Medical complications of diabetes include microangiopathy, cataracts, blindness due to retinopathy, neuropathy, decreased resistance to infections, and hyperlipidemia.

CAPILLARY EFFECTS

The capillary walls in diabetic patients thicken due to collagen and glycoprotein deposits. The thickened capillaries are less flexible and more susceptible to blockage and atherosclerosis. Rat aorta smooth muscle cells incorporated less radio-labeled amino acids when cultured with *V. myrtillus* anthocyanins (VMA) (Boniface et al., 1986), suggesting one mechanism by which bilberry maintains normal capillary structure. Similar results were obtained for collagen content in cells from diabetic rats, and VMA reduced collagen content more than did insulin.

Boniface and colleagues (1986) also reported positive results for diabetic patients treated with 500–600 mg of VMA for 8–33 months. Improvements in capillaries and microaneurysms were found for many, but not all, patients. Although there was no change in soluble collagen due to VMA, insoluble collagen levels returned to levels similar to those in normal subjects, and structural glycoproteins in diabetics were reduced by 30%. Based upon this series of experiments, the French scientists recommended that diabetic patients be given 500 mg VMA daily, split in 2 doses, for a period of at least several months. Sporadic doses of lower quantities were not believed to be of value.

GLYCEMIA

An aqueous alcohol extract of *V. myrtillus* leaves produced a 26% reduction in plasma glucose levels in streptozotocin-induced diabetic rats (Cignarella et al., 1996). Plasma triglycerides decreased in proportion with the amount of bilberry leaf extract given to rats (1.2 or 3.0 g/kg body weight) fed a hyperlipidemic diet, but the reduction was less than that obtained for rats treated with ciprofibrate (10 or 20 mg/kg). Ciprofibrate also reduced plasma levels of free fatty acids, the precursors for triglycerides, while leaf extract affected only triglycerides. Increased tendency to form blood clots or thrombi represents another diabetes-related risk for cardiovascular problems. Thrombus weight and protein content was lower in diabetic rats given ciprofibrate, but not for those rats given bilberry leaf extract.

Flavonoids interfere with starch enzymes and therefore affect post-prandial blood glucose levels because glucose is released gradually in the small intestine. An *in vitro* survey of plant materials for ß-glucosidase inhibitory activity found no activity in blueberry (species not identified) skin extract (Matsui et al., 2001a), presumably because acylated anthocyanins in other plants have greater inhibitory ability (4–18 mM) (Matsui et al., 2001b). Pectin and other types of soluble fiber in berries also slow digestion of carbohydrates and subsequent absorption of sugars. An *in vitro* study indicated that fiber affects glucose levels by three mechanisms: increased intestinal content viscosity that slows diffusion to the brush border, entrapment of glucose, and physical separation of enzyme and substrate (Ou et al., 2001).

Flavonoids inhibit aldose reductase, which converts glucose to sorbitol (Varma, 1986). This conversion is associated with many of the adverse outcomes associated with diabetes. We have demonstrated that anthocyanins and anthocyanin-rich fruit products can inhibit recombinant aldose reductase *in vitro* (Figure 3.3). Enzyme inhibition by juice powders was not highly correlated with anthocyanin content, suggesting that ascorbic acid and other antioxidants in fruit contribute to inhibition.

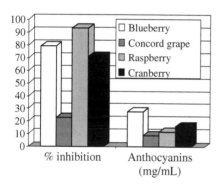

Figure 3.3 *In vitro* aldose reductase inhibition and anthocyanin content of spray-dried fruit powders.

AGING AND MEMORY LOSS

OXIDATIVE STRESS

As people age, balance, coordination, short-term memory, information retrieval, and other brain tasks are hindered. A lifetime of oxidative stress may induce damages to lipids and proteins that lead to cellular damage. Dementia results from impaired vascular supply, both chronic and acute, and from the formation of amyloid plaques associated with Alzheimer's disease. An animal study recently provided some evidence that dietary antioxidants can protect the brain from oxygen-induced damage (Joseph et al., 1998). Six- to eight-month-old F344 rats were fed a control diet or diets containing antioxidants (500 IU/kg vitamin E, 10 g/kg dried aqueous blueberry extract, 9.4 g/kg dried strawberry extract, or 6.7 g/kg dried spinach extract). After 8 weeks of the diet, the rats were subjected to 48 h of 100% O_2 to induce damage similar to that found in aged rats. All antioxidant diets prevented decreases in nerve growth factor in the basal forebrain, and other adverse effects were ameliorated. In another study, 19-month-old Fisher 344 rats fed a blueberry diet (14.8 g dried aqueous extract per kg) generally performed better in cognitive performance than did rats fed strawberry or spinach diets, but only the blueberry-fed rats demonstrated a reversal in age-related motor deficits (Joseph et al., 1999).

Differences in polyphenolic composition were thought to be responsible for varying effects of rabbiteye (cv. Tif) and lowbush blueberry extracts fed to rats for 8 weeks (Youdim et al., 2000). Rats fed either type of berry had higher striatal concentrations of dopamine and ascorbic acid than did controls. Psychomotor skills varied, but for all tests at least one blueberry group had superior results compared with rats fed the control diet.

HORMONAL INFLUENCES

The thyroid hormones thyroxine (T4) and triiodothyronine (T3) regulate temperature and metabolism. In rats given intraperitoneal injections of bilberry anthocyanins (200 mg/kg/day) for 5 days, significantly more T3 was found in their brains than rats given only the solvent (26% alcohol) (Saija et al., 1990). T3 enters the brain via specific transport in the capillaries, therefore anthocyanins may mediate T3 transport at the capillary level. The specific portions of the brain that contained more T3 in the bilberry-treated animals were those responsible for memory, vision, and control of sensory input — the frontal, temporoparietal and occipital cortexes, and the components of the limbic system (hippocampus, thalamus, and hypothalamus).

VISION

Bilberry extract is prescribed in Europe for eyesight, particularly night vision. This health benefit is the primary reason for the product's popularity in Japan and Korea, where it is used to relieve computer-induced eyestrain (Kalt and Dufour, 1997). This benefit may be due, in part, to the small amounts of carotenoids present in the fruit. However, a double-blind, placebo-controlled study demonstrated that oral doses of anthocyanins are important for regeneration of visual purple (Alfieri and Sole, 1966). Bilberry extracts appear to benefit vision in several ways: improved night vision by enhanced regeneration of retinal pigments, increased circulation within the capillaries of the retina, inhibition of Maillard reactions in the lens to reduce cataract formation, and protection from ultraviolet light. The antioxidant properties of V. myrtillus extracts may be responsible for these health benefits. Antioxidants have been suggested to retard oxidation in the lens and to slow retinal angiopathy, which occurs in both age-related macular degeneration and diabetic retinopathy (Trevithick and Mitton, 1999).

Using adapto-electroretinograms (AERG) to assess dark adaptation, six subjects adapted to light within 6.5 minutes, compared with 9 minutes for the control group.

A dose of 500 mg *V. myrtillus* anthocyanins with 25 mg ß-carotene improved vision in poor light due to improved mesoptic vision and reduced sensibility at direct glare (Scialdone, 1966). Bilberry extract slightly improved retinal sensitivity in 16 normal adults (Magnasco, 1966). In a study of 20 25-year-olds with good vision, the group (N = 30) receiving bilberry extract had slightly, but significantly, improved adaption related to night vision compared to the placebo group (Jayle et al., 1965).

Inhibition of aldose reductase by *Vaccinium* flavonoids should aid in minimizing lens damage in diabetics since free sugars form Maillard-type products that contribute to lens clouding. Quercetin decreased sugars in sugar alcohols in the lens of diabetic degus; quercitrin inhibited xylitol synthesis in rat lens (Varma, 1986). Diabetic hyperopia was controlled in rabbits given flavonoids before or after administration of alloxan to induce diabetes.

No studies have been published to date that describe the use of blueberry or cranberry products to treat visual problems, although these products are being used as dietary supplements for this purpose. The dose of anthocyanins required to improve or maintain vision is not yet fully understood, and the lower concentration of the pigments in blueberries suggest that they may be less effective. During the spring of 2002 we will conduct a study involving 100 middle-aged adults who will be given dried blueberry juice or a placebo for 3 months, with periodic assessment of contrast sensitivity.

ANTICANCER ACTIVITY

Consumption of fruit is linked to reduced risks for several cancers. Research involving *Vaccinium* species has focused on *in vitro* and animal studies. The relative value of fruit components is not clear; thus, estimation of recommended consumption levels is difficult. Spray-dried *V. angustifolium* juice was fractionated for several *in vitro* assays (Smith et al., 2000). A crude 70% acetone extract induced quinone oxidase, suggesting potential benefits in inhibiting carcinogenesis. Other studies are summarized in Table 3.1.

URINARY TRACT HEALTH

Bilberries, blueberries, and cranberries have been used in several countries to treat urinary and digestive tract infections. In the 1960s, several studies examined the effects of anthocyanins on pathogenic bacteria. Monoglucosides of delphinidin, malvidin, and petunidin varied in their ability to inhibit key microbial enzymes (Carpenter et al., 1967). Aflatoxin B_1 production by *Aspergillus flavus* was inhibited by anthocyanidins in the following order: delphinidin > pelargonidin > peonidin > cyanidin (Norton, 1999). Glycosides were less effective in toxin suppression than the aglycones. It is not known whether these compounds retard toxins formed by mold species found on *Vaccinium* fruit.

TABLE 3.1
Summary of Research on Anticancer Properties of *Vaccinium* Fruits.

Fruit product	Results	Reference
Blueberry juice	Decreased mutagenicity of polycyclic aromatic hydrocarbons in the Ames assay	Edenharder et al., 1994
Solvent fractions of *V. angustifolium, myrtilis, macrocarpon,* and *vitis-idaea*	Proanthocyanidins and crude extracts inhibit ornithine decarboxylase activity at lower IC50 than did ethyl acetate extracts and anthocyanins	Bomser et al., 1996
Anthocyanins	Aglycons decreased growth of the human vulva carcinoma cell line A431; glycosides were effective at concentrations > 100 μM	Meiers et al., 2001

URINARY TRACT INFECTIONS

According to the National Institute of Diabetes and Digestive and Kidney Diseases (2001), urinary tract infections (UTI) resulted in over 8.5 million doctor visits in 1997. These infections are painful and can lead to serious infections of other organs, usually the kidneys. Interstitial cystitis afflicts about 500,000 Americans annually; 90% of these sufferers are women. As much as half of the older women in long-term care facilities or hospitals may suffer from UTIs (Monane, et al.1995). UTIs tend to recur, but statistics on the recurrence rate are not available. Prevention of these infections can significantly reduce medical expenses and improve quality of life for persons prone to recurrent infections.

UTIs develop when bacteria, and occasionally yeasts and fungi, are introduced into the normally sterile urinary tract, which consists of the bladder and ureters. Microbes may enter the tract in a number of ways, primarily from sexual intercourse, poor hygiene, and catheters. Painful and frequent urination, burning, and fever result when the bacteria multiply. Conventional medical therapy consists of antibiotic administration along with increased consumption of water. The additional liquid is believed to aid in flushing bacteria from the tissues. In the past few years, the medical community and the public have become concerned about the development of antibiotic-resistant strains of bacteria. Recurrent UTIs treated with the same antibiotic could lead to selection for drug-resistant bacteria. Without an effective antibiotic, such bacteria could proliferate and spread to the kidneys and other vital organs.

Concerns about antibiotics and other medications have spurred more Americans to seek alternative, "natural" treatments for ailments. One widely accepted concept is the use of cranberry juice cocktail to prevent or treat UTIs. Although the earliest history of this practice is not known, many people assume that Native Americans discovered the beneficial properties of cranberries (*V. macrocarpon*) before the arrival of European colonists. The first scientific study of the medicinal properties of cranberry products was published in 1923. Nearly 80 years later, the medical community is still divided over the actual benefits of cranberry juice for UTIs.

Mechanisms for Reducing UTI

Bacteria can be killed in a number of ways: extremes of temperature, radiation, electrical currents, extremes of pH, and numerous antimicrobial compounds. Obviously the first three options are not possible within a patient, but most fruit juices have a low pH (high acidity). Many juices also contain organic acids and other compounds that naturally inhibit bacteria. *Escherichia coli* are gram-positive bacteria naturally present in the large intestine, but they may contaminate the urinary tract. *E. coli* have pili or fimbriae on their surface that resemble hairs. These fimbriae contain two types of adhesins, compounds that promote adhesion to other cells, such as the epithelial cells lining the bladder. One adhesin is called mannose-specific (MS) since it preferentially binds to the sugar mannose. Fructose is the only other sugar that can bind to this adhesin. Most *E. coli* strains from urinary tract infections have Type I fimbriae that have this MS adhesin.

The other adhesin is referred to as mannose-resistant (MR), since it does not bind to either sugar. MR adhesin is found in *E. coli* strains from both diarrhea and urinary tract infections. This adhesin appears to be inhibited by large-molecular weight compounds that are not yet clearly identified (Ofek et al., 1996). Binding these inhibitors prevents the bacteria from attaching to other cells, thus producing an inhibitory effect on their infective ability.

Blueberries and cranberries appear to have unique phytochemical mixtures that aid in reduction of urinary tract pathogens. The combination of low pH, organic acids, fructose, and MR-adhesin inhibitors provide a hurdle effect (Figure 3.4). By using several methods to kill bacteria, or in the case of UTIs, preventing adhesion of bacteria to tissues, bacterial strains resistant to one process will be overcome by another, resulting in a minimal microbial risk.

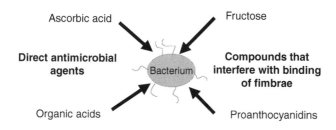

Figure 3.4 Possible mechanisms for antimicrobial activity of *Vaccinium* phytochemicals.

Urine Acidification

The first theory proposed to explain why drinking cranberry juice seemed to help some patients was based on the idea that the naturally tart cranberry juice contained acids that lowered the pH of urine. Normal urine pH is close to neutral (7.0); a pH of 5.5 or lower is necessary to retard development of UTIs. An early clinical study found that 53% of 60 patients with acute UTIs showed some improvement after consuming 16 ounces of cranberry juice daily for three weeks; 27% had no change in symptoms (Papas et al., 1966). The researchers suggested that cranberry juice aided the drug methenamine by keeping urine pH sufficiently low for the drug to have enhanced antibacterial activity. They concluded that cranberry juice was a helpful, well-tolerated, and inexpensive adjunct to traditional antibiotic therapy.

Sweetened, diluted cranberry juice (about 80% juice) was provided along with meals to 40 healthy young adults (Kinney and Blount, 1979). Each treatment group (150, 180, 210, or 240 mL cranberry juice per meal) had significantly lower urine pH than the control group. There was no apparent dose relationship; i.e., it did not appear to matter how much juice each person drank. However, incidence of urinary tract infections was not monitored, and it is unlikely that most people would drink a product containing 80% cranberry juice, even if sweetened, due to the strong bitter and sour flavor. In a double-crossover experiment with 21 elderly men, no cranberry juice cocktail was provided for 4 weeks, followed by 4 weeks in which two 118.3 mL servings were consumed daily; then the experiment concluded with another 4-week period with no cranberry juice (Jackson and Hicks, 1997). Urine pH was significantly lower during the juice phase of the program.

Organic Acids

Fellers and coworkers (1933) first proposed that cranberry juice might retard bacteria via formation of hippuric acid from organic acids found in the berries. Benzoic and quinic acids are converted to hippuric acid in order for the body to more easily excrete them in the urine. Hippuric acid acts directly upon bacteria by causing electrical imbalances that result in cell death. Benzoic acid itself inhibits bacteria when it is present at concentrations of 0.1–0.2%.

Inhibition of Bacterial Adhesins

As mentioned earlier, bacteria causing UTIs have one of two types of adhesins that allow the bacteria to stick to tissues in the body. Urine from mice and humans fed cranberry juice cocktail prevented adhesion of *E. coli* to isolated bladder cells (Sobota, 1984), lending support to the idea that active compounds must be present in the urine. The same study diluted cranberry juice, cranberry juice cocktail, and concentrate, and dilution up to 1:100 significantly reduced bacterial adhesion compared with the saline control. A subsequent study found that cranberry juice was less effective in removing bacteria that were already attached to tissues (Schmidt and Sobota, 1988), suggesting that cranberry juice may have more potential as a preventative measure than as a cure. Drinking cranberry juice twice daily reduced the number of blood cells in the urine of 17 disabled children, but *E. coli* was still recovered (Rogers, 1991). Cranberry juice appeared to reduce mucus formation in those catheterized children with spina bifida.

Yeast aggregation by *E. coli* is used to measure the mannose-binding potential of the bacteria; this test approximates the ability of bacteria to adhere to urinary tract tissue. Fructose concentrations of 5%, whether in cranberry juice cocktail or as a fructose solution, produced 50% inhibition yeast agglutination by three strains of *E. coli*, even at a 1:52 dilution for Strain 346 (Zafriri, et al., 1989).

A letter to the editor of the *New England Journal of Medicine* reported that only blueberry and cranberry juices contained an MR-inhibitor among seven juices tested (Ofek et al., 1991). However, only data for the cranberry inhibitor has been published. Correspondence with Dr. Itzhak Ofek of Tel Aviv University revealed that the group had little data on blueberries, and that no data had been published on the inhibitory effect of blueberries. The inhibitor from both cranberries and blueberries had a molecular weight over 15000, and was acid-resistant but sensitive to high pH (Ofek et al., 1996).

Clinical Studies

A highly publicized study examined the effect of cranberry juice cocktail on bacteriuria and pyuria in 153 elderly women (Avorn et al., 1994). By the second month of the study, the placebo group had three times the number of bacteria-contaminated urine samples as the cranberry group. Although the differences varied with time, the differences between groups became smaller after 6 months. Some patients in each group (16, placebo; 8, cranberry) were treated with antibiotics for UTIs during the course of the study. A major critique of this study is that the placebo group had a much greater number of patients who had been treated in the past 6 months for UTIs. This group may have been predisposed to development of such infections, thereby skewing the results in favor of the cranberry treatment. However, the researchers did account for the difference in groups with statistical analyses. Interestingly, all cranberry beverage samples inhibited bacterial adhesion *in vitro*, while none of the placebo beverages inhibited adhesion.

In a recent study involving a blend of juices from European cranberry (*V. oxycoccus*) and lingonberry, 150 women were assigned to drink the juice, a probiotic beverage, or no treatment for 6–12 months, depending upon the treatment (Kontiokari et al., 2001). Although the study was not placebo controlled, the juice significantly delayed the onset of UTI recurrence (Figure 3.5).

Although most studies have reported the effect of beverages, capsules containing 400 mg of cranberry solids were tested in a small study (Walker et al., 1997). Only ten women completed the study, which involved consumption of the cranberry supplement or a calcium placebo for 6 months after diagnosis and treatment with antibiotics of a UTI. The researchers concluded that the cranberry capsules were effective in reducing development of UTIs in young women prone to such infections.

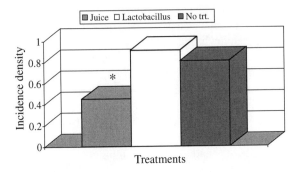

Figure 3.5 Incidence density of urinary tract recurrence in 150 women in 3 treatment groups (N = 50).* indicates significant difference (p = 0.03). (Source: Kontiokari et al. 2001. *Brit. Med. J.,* 322(7302):1571–1575. With permission.)

A review article for nurses summarized situations in which cranberry juice cocktail could be used: as a preventative measure for persons prone to UTIs, as part of the treatment for UTIs, to acidify urine to prevent calcium stone formation and improve urine odor, and to reduce mucus formation in patients with urostomies or catheters (Leaver, 1996). Fleet (1994) suggested that the effect of cranberry juice on intestinal flora may be an important first step in reducing UTIs, since bladder contamination with fecal bacteria is common. Significantly fewer bacteria were found adhering to bladder cells from 15 persons with spinal cord injuries who had drunk three glasses of cranberry juice (Reid et al., 2001). Howe and Bates (1987) questioned the high cost of cranberry juice and raised the issue of taste. Persons who dislike the taste of cranberry juice will be reluctant to drink it, despite potential health benefits. Table 3.2 lists other conditions that may be improved by cranberry or blueberry consumption.

ISSUES TO BE RESOLVED

Opinions differ on how and if these compounds enter the body. Glycosides could be cleaved at the intestinal brush border or by microbes, causing only aglycones to

TABLE 3.2
Potential Antimicrobial Applications for *Vaccinium* Products

Condition	Reference
Dental plaque	Weiss et al. 1998
Stomach ulcer	Burger et al., 2000
Diarrhea	Folklore

be absorbed. However, intact anthocyanins may be absorbed. Published studies of anthocyanin absorption have used pure compounds or extracts, not whole foods, from which extraction and absorption is presumably more difficult. *V. myrtillus* anthocyanins given to rats by either oral or intravenous routes were recovered from urine and bile (Lietti and Forni, 1976b). Ring fission products were isolated from the urine of rats fed either delphinidin or malvin (malvidin-3,5-diglucoside) (Griffiths and Smith, 1972).

Improved analytical techniques have enabled researchers to better identify phytochemicals in biological samples. Direct feeding of cyanidin-3-glucoside and cyanidin-3,5-diglucoside resulted in recovery of the compounds and their metabolites from both rat and human plasma (Miyazawa et al., 1999). In four healthy elderly women fed 12 g of elderberry extract, which contained 720 mg anthocyanins, peak plasma anthocyanins levels (97.4 nmol/L) were achieved 71 min after consumption (Cao et al., 2001). The estimated anthocyanin half-life was 132.6 min, and both glycosides and aglycones were identified in the women's urine. The fate of these compounds is not clear. Health effects persist long after the presumed clearance; perhaps these compounds adhere to capillaries and other tissues.

Even less is known about the absorption and metabolism of larger compounds such as the proanthocyanidins. Mice fed cranberry juice cocktail or isolated proanthocyanidins (50 or 500 mg/300mL water) excreted urine with *E. coli* anti-adherence activity *in vitro* (Howell et al., 2001). The researchers concluded that these compounds must be absorbed in order to have reached the urinary tract.

DESIGN OF CLINICAL STUDIES

It is not clear what constitutes a typical dose of these products. The blueberry industry recommends 1/2 cup of fresh or frozen berries daily; cranberry researchers are urged to test the equivalent of one 8 oz glass of cranberry juice cocktail (with 27–31% cranberry juice) per day. Another issue is the use of foods vs. supplements or purified compounds in studies. Use of berries or juice could cause compliance problems, and certainly makes it difficult to design placebos. Arguments against purified or processed materials include loss of bioactivity in processing, and lack of interaction among multiple phytochemicals and dietary fiber in the whole berries. A final consideration is whether the study goal is to prevent or cure a health condition. Larger doses maybe needed for longer periods, and effective dose levels must be determined.

The fruits of the *Vaccinium* family contain a variety of phytochemicals with diverse health benefits. Publicity surrounding these fruits has spurred more researchers to examine the effects of these fruits and their components. While sufficient evidence does not yet exist for health claims, structure-function claims are possible for dietary supplements. Fortunately these berries are available in appealing products with low toxicity; thus, consumers may be willing to consume them before research validates potential benefits.

REFERENCES

Alfieri, R., and Sole, P. 1966. Influence des anthocyanosides administrés par voie oro-perlinguale sur l'adapto-électrorétinogramme (AERG) en lumiére rouge chez l' Homme, *Comptes Rendus des Séances de la Société de Biologie et des Ses Filiales,* 160(8):1590-1593.

Amouretti, M. 1972. Intérêt thérapeutique des anthocyanosides de Vaccinium myrtillus dans un service de médecine interne, *Thérapeutique,* 48(9):579–581.

Andriambeloson, E., Magnier, C., Haan-Archipoff, G., Lobstein, A., Anton, R., Beretz, A., Stoclet, J.C., and Andriantsitohaina, R. 1998. Natural dietary polyphenolic compounds cause endothelium-dependent vasorelaxation in rat thoracic aorta, *J. Nutr.,* 128(12):2324–2333.

Andriambeloson, E., Stoclet, J.C., and Andriantsitohaina, R. 1996. Effets vasculaires d'extraits végétaux contenant des dérivés polyphénoliques, in *Polyphenols Communications 96,* J. Vercauteren, C. Chèze, M.C. Dumon, and J.F. Weber, 2:421–422. Groupe Polyphénols, Bordeaux, France.

Avorn, J., Monane, M., Gurwitz, J.H., Glynn, R.J., Choodnovskiy, I., and Lipsitz, L.A., 1994. Reduction of bacteriuria and pyuria after ingestion of cranberry juice, *JAMA,* 271(10):751-754.

Bertuglia, S., Malandrino, S., and Colantuoni, A. 1995. Effect of *Vaccinium myrtillus* anthocyanosides on ischaemia reperfusion injury in hamster cheek pouch microcirculation, *Pharmacol. Res.,* 31(3/4):183–187.

Bettini, V., Fiori, A., Martino, R., Mayellaro, R., and Ton, P. 1985. Study of the mechanism whereby anthocyanosides potentiate the effect of catecholamines on coronary vessels, *Fitoterapia,* 56(2):67–72.

Bettini, V., Martino, R., Tegazzin, V., and Ton, P. 1987. Risposte contrattili di segmenti di arterie coronarie all'istamina e all'angiotensina II in presenza di antocianosidi del mirtillo, *Cardiologia,* 32(10):1155–1159.

Boger, R.H., Lentz, S.R., Bode-Boger, S.M., Knapp, H.R., and Haynes, W.G. 2001. Elevation of asymmetrical dimethylarginine may mediate endothelial dysfunction during experimental hyperhomocyst(e)inaemia in humans, *Clin. Sci.,* (Colch), 100(2):161–167.

Bomser, J., Madhavi, D.L., Singletary, K., and Smith, M.A.L. 1996. *In vitro* anticancer activity of fruit extracts from *Vaccinium* species, *Planta Med.,* 62(3):212–216.

Boniface, R., Miskulin, M., Robert, L., and Robert, A.M. 1986. Pharmacological properties of *Myrtillus* anthocyanosides: correlation with results of treatment of diabetic microangiopathy, in *Flavonoids and Bioflavonoids*, 1985, L. Farkas, M. Gabor, and F. Kallay, eds., Amsterdam: Elsevier, pp. 293–301.

Burger, O., Ofek, I., Tabak, M., Weiss, E.I., Sharon, N., and Neeman, I. 2000. A high molecular mass constituent of cranberry juice inhibits helicobacter pylori adhesion to human gastric mucus, *FEMS Immunol. Med. Microbiol.,* 29(4):295–301.

Cao, G., Muccitelli, H.U., Sánchez-Moreno, C., and Prior, R.L. 2001. Anthocyanins are absorbed in glycated forms in elderly women: a pharmacokinetic study, *Am. J. Clin. Nutr.,* 73(4):920–926.

Cao, G., Martin, A., Sofic, E., McEwen, J., O'Brien, C., Lischner, N., Ehlenfeldt, M., Kalt, W., Krewer, G., and Mainland, C.M. 1998. Antioxidant capacity as influenced by total phenolic and anthocyanin content, maturity, and variety of *Vaccinium* species, *J. Agric. Food Chem.,* 46(7):2686–2693.

Carpenter, J.A., Wang, Y.-P., and Powers, J.J. 1967. Effect of anthocyanin pigments on certain enzymes, *Proc. Soc. Exp. Biol. Med.*, 124(3):702–706.

Chen, H., Zuo, Y., and Deng, Y. 2001. Separation and determination of flavonoids and other phenolic compounds in cranberry juice by high-performance liquid chromatography, *J. Chromatogr. A.,* 913(1-2):387–395.

Cignarella, A., Nastasi, M., Cavalli, E., and Puglisi, L. 1996. Novel lipid-lowering properties of *Vaccinium myrtillus* L. leaves, a traditional antidiabetic treatment, in several models of rat dyslipidaemia: a comparison with ciprofibrate, *Thrombosis Res.,* 84(5):311–322.

Coget, J. and Merlen, J.F. 1968. Étude clinique d'un nouvel agent de protection vasculaire le Difrarel 20, composé d'anthocyanosides extrait du vaccinum myrtillus, *Phlâebologie,* 21(2):221–228.

Costantino, L., Albasini, A., Rastelli, G., and Benvenuti, S. 1992. "Activity of polyphenolic crude extracts as scavengers of superoxide radicals and inhibitors or xanthine oxidase," *Planta Medica,* 58:342–344.

Cristoni, A. and Magistretti, M.J. 1987. Antiulcer and healing activity of *Vaccinium myrtillus* anthocyanosides, *Farmaco,* 42(2):29–43.

Eck, P. 1988. *Blueberry Science*, Rutgers Univ. Press, New Brunswick, NJ.

Edenharder, R., Kurz, P., John, K., Burgard, S., and Seeger, K. 1994. *In vitro* effect of vegetable and fruit juices on the mutagenicity of 2-amino-3-methylimidazo[4,5-*f*]quinoline, 2-amino-3,4-dimethylimidazo[4,5-*f*]quinoline and 2-amino-3,8-dimethylimidazo[4,5-*f*]quinoxaline, *Food Chem. Toxicol.,* 32(5):443–459.

Fellers, C.R., Redmon, B.C., and Parrott, E.M. 1933. Effect of cranberries on urinary acidity and blood alkali reserve, *J. Nutr.,* 6:455-463.

Fitzpatrick, D.F., Hirschfield, S.L., Ricci, T., Jantzen, P., and Coffey, R.G. 1995. Endothelium-dependent vasorelaxation caused by various plant extracts, *J. Cardiovascular Pharmacol.,* 26(1):90-95.

Fleet, J.C., 1994. New suppport for a folk remedy: cranberry juice reduces bacteriuria and pyuria in elderly women, *Nutr. Rev.,* 52(5):168-178.

Griffiths, L.A. and Smith, G.E. 1972. Metabolism of myricetin and related compounds in the rat metabolite formation *in vivo* and by the intestinal microflora *in vitro, Biochem. J.,* 130:141–151.

Heinonen, I.M., Meyer, A.S., and Frankel, E.N. 1998. Antioxidant activity of berry phenolics on human low-density lipoprotein and liposome oxidation, *J. Agric. Food Chem.,* 46(10):4107–4112.

Howe, S.M. and Bates, P., 1987. The cranberry juice cure: fact or fiction, *AUAA J.,* July-September pp. 13, 16.

Howell, A.B., Leahy, M., Kurawska, E., and Guthrie, N. 2001. *In vivo* evidence that cranberry proanthocyanidins inhibit adherence of p-fimbriated *E. coli* bacteria to uroepithelial cells, Abstract 244.1, *FASEB J.,* 15(4):A284.

Jackson, B. and Hicks, L.E. 1997. Effect of cranberry juice on urinary pH in older adults, *Home Healthcare Nurse,* 15(3):199–202.

Jayle, G.E., Aubry, M., Gavini, H., Braccini, G., and de la Baume, C. 1965. Étude concernant l'action sur la vision nocturne, *Ann. Oculistique,* 198(6):556–562.

Jonadet, M., Meunier, M.T., Bastide, J., and Bastide, P. 1983. Anthocyanosides extraits de *Vitis vinifera*, de *Vaccinium myrtillus* et de *Pinus maritimus*. I. Activités inhibitrices vis-á-vis de l'élastase *in vitro*; II. Activités angioprotectrices comparées *in vivo, J. Pharm. Belg.,* 38(1):41–46.

Joseph, J.A., Denisova, N., Fisher, D., Shukitt-Hale, B., Prior, R., and Cao, G. 1998. Membrane and receptor modifications of oxidative stress vulnerability in aging. Nutritional considerations, *Ann. NY Acad. Sci.,* 854(11):268–276.

Joseph, J.A., Shukitt-Hale, B., Denisova, N.A., Bielinski, D., Martin, A., McEwen, J.J., and Bickford, P.C. 1999. Reversals in age-related declines in neuronal signal transduction, cognitive, and motor behavioral deficits with blueberry, spinach, or strawberry dietary supplementation. *J. Neuroscience,* 19(18):8114–8121.

Kader, F., Rovel, B., Giradin, M., and Metche, M. 1996. Fractionation and identification of the phenolic compounds of highbush blueberries (*Vaccinium corymbosum*, L.), *Food Chem.,* 55(1):35–40.

Kalt, W. and Dufour, D. 1997. Health functionality of blueberries, *HortTechnol.,* 7(3):216–222.

Kinney, A.B. and Blount, M. 1979. Effect of cranberry juice on urinary pH, *Nursing Res.,* 28(5):287–290.

Kontiokari, T., Sundqvist, K., Nuutinen, M., Pokka, T., Koskela, M., and Uhari, M. 2001. Randomised trial of cranberry-lingonberry juice and Lactobacillus GG drink for the prevention of urinary tract infections in women, *Brit. Med. J.,* 322(7302):1571–1575.

Laplaud, P.M., Lelubre, A., and Chapman, M.J. 1997. Antioxidant action of *Vaccinium mytrillus* extract on human low density lipoproteins *in vitro*: initial observations, *Fund. Clin. Pharmacol.,* 11(1):35–40.

Leaver, R.B., 1996. Cranberry juice: an update on cranberry juice, focusing on its use in the prevention and treatment of urinary tract infection, its effect on the formation of urinary stones and on the control of mucus formation, *Prof. Nurse,* 11(8):525-526.

Lietti, A., Cristoni, A., and Picci, M. 1976a. Studies on *Vaccinium myrtillus* anthocyanosides. I. Vasoprotective and anti-inflammatory activity, *Arzneim.-Forsch.,* 26(5):829–832

Lietti, A. and Forni, G. 1976b. Studies on *Vaccinium myrtillus* anthocyanosides. II. Aspects of anthocyanins pharmacokinetics in the rat, *Arzneim.-Forsch.,* 26(5):832–835.

Magnasco, A. 1966. Influenza degli antocianosidi sulla soglia retinica differenziale mesopica, *Ann. Ottalmologia Clin. Oculistica,* 92(3):188-193.

Marniemi, J., Hakala, P., Maki, J., and Ahotupa, M. 2000. Partial resistance of low density lipoprotein to oxidation *in vivo* after increased intake of berries, *Nutr. Metab. Cardiovasc. Dis.,* 10(6):331–337.

Matsui, T., Ueda, T., Oki, T., Sugita, K., Terahara, N., and Matsumoto, K. 2001a. α-Glucosidase inhibitory action of natural acylated anthocyanins. 1. Survey of natural pigments with potent inhibitory activity, *J. Agric. Food Chem.,* 49(4):1948–1951.

Matsui,T., Ueda, T., Oki, T., Sugita, K., Terahara, N., and Matsumoto, K. 2001b. α-Glucosidase inhibitory action of natural acylated anthocyanins. 2. α-Glucosidase inhibition by isolated acylated anthocyanins, *J. Agric. Food Chem.,* 49(4):1952–1956.

Mazza, G., and Miniati, E. 1993. *Anthocyanins in Fruits, Vegetables, and Grains*, CRC Press, Inc., Boca Raton, FL.

Meiers, S., Kemeny, M., Weyand, U., Gastpar, R., von Angerer, E., and Marko, D. 2001. The anthocyanidins cyanidin and delphinidin are potent inhibitors of the epidermal growth-factor receptor, *J. Agric. Food Chem.,* 49(2):958–962.

Meyer, A.S., Heinonen, M., and Frankel, E.N. 1998. Antioxidant interactions of catechin, cyanidin, caffeic acid, quercetin, and ellagic acid on human LDL oxidation, *Food Chem.,* 61(1):71–75.

Mian, E., Curri, S.B., Lietti, A., and Bombardelli, E. 1977. Anthocyanosides and the walls of the microvessels: further aspects of the mechanism of action of their protective effect in syndromes due to abnormal capillary fragility, *Minerva Med.,* 68(52):3565–3581.

Miyazawa, T., Nakagawa, K., Kudo, M., Muraishi, K., and Someya, K. 1999. Direct intestinal absorption of red fruit anthocyanins, cyanidin-3-glucoside and cyanidin-3,5-diglucoside, into rats and humans, *J. Agric. Food Chem.,* 47(3):1083–1091.

Monane, M., Gurwitz, J., Lipsitz, L., Glynn, R., Choodnovskiy, I., and Avorn, J. 1995. Epidemiologic and diagnostic aspects of bacteriuria: a longitudinal study in older women, *J. Am. Geriatric Soc.*, 43(6):618–622.

Moore, J.N. 1994. The blueberry industry of North America, *HortTechol.*, 4(2):96-102.

Norton, R.A. 1999. Inhibition of aflatoxin B$_1$ biosynthesis in *Aspergillus flavus* by anthocyaninidins and related flavonoids, *J. Agric Food Chem.*, 47(3):1230–1235.

Ofek, I., Goldhar, J., and Sharon, N. 1996. Anti-*Escherichia coli* adhesin activity of cranberry and blueberry juices, in *Toward Anti-Adhesion Therapy for Microbial Diseases*, Kahane and Ofek, eds., Plenum Press, New York, pp. 179–183.

Ofek, I., Goldhar, J., Zafriri, D., Lis, H., Adar, R, and Sharon, N. 1991. Anti-*Escherichia coli* adhesin activity of cranberry and blueberry juices, *New Engl. J. Med.*, 324(22):1599.

Ohnishi, M., Morishita, H., Iwahashi, H., Toda, S., Shirataki, Y., Kimura, M., and Kido, R. 1994. Inhibitory effects of chlorogenic acids on linoleic acid peroxidation and haemolysis, *Phytochem.*, 36(3):579–583.

Ou, S., Kwok, K.-C., Li, Y., and Fu, L. 2001. *In vitro* study of possible role of dietary fiber in lowering postprandial serum glucose, *J. Agric. Food Chem.*, 49(2):1026–1029.

Papas, P. N. Brusch, C.A. and Cresia, G.C., 1966. Cranberry juice in the treatment of urinary tract infections, *Southwestern Med.*, 47(1):17-20.

Pedersen, C.B., Kyle, J., Jenkinson, A.M., Gardner, P.T., McPhail, D.B., and Duthie, G.G. 2000. Effects of blueberry and cranberry juice consumption on the plasma antioxidant capacity of healthy female volunteers, *Eur. J. Clin. Nutr.*, 54(5):405–08.

Reid, G., Hsiehl, J., Potter, P., Mighton, J., Lam, D., Warren, D., and Stephenson, J. 2001. Cranberry juice consumption may reduce biofilms on uroepithelial cells: pilot study in spinal cord injured patients, *Spinal Cord*, 39(1):26–30.

Rice-Evans, C.A., Miller, N.J., Bolwell, P.G., Bramley, P.M., and Pridham, J.B. 1995. The relative antioxidant activities of plant-derived polyphenolic flavonoids, *Free Rad. Res.*, 22(4):375–383.

Rogers, J., 1991. Pass the cranberry juice, *Nursing Times*, 87(48):36-37.

Saija, A., Princi, P., D'Amico, N., DePasquale, R., and Costa, G., 1990. Effect of *Vaccinium myrtillus* anthocyanins on triiodothyronine transport into brain in the rat, *Pharmacol. Res.*, 22(1):59–60.

Satué-Gracia, M.T., Heinonen, M., and Frankel, E.N. 1997. Anthocyanins as antioxidants on human low-density lipoprotein and lecithin-liposome systems, *J. Agric. Food Chem.*, 45(9):3362–3367.

Scialdone, D. 1966. L'azione delle antocianine sul senso luminoso, *Ann. Ottalmologia clin. oculistica*, 92(1):43–51.

Schmidt, D.R. and Sobota, A.E. 1988. An examination of the anti-adherence activity of cranberry juice on urinary and nonurinary bacterial isolates, *Microbios*, 55(224-225):173–181.

Schramm, D.D. and J.B. German. 1998. Potential effects of flavonoids on the etiology of vascular disease, *J. Nutr. Biochem.*, 9(10):560–566.

Smith, M.A.L., Marley, K.A., Seigler, D., Singletary, K.W., and Meline, B. 2000. Bioactive properties of wild blueberry fruits, *J. Food Sci.*, 65(2):352–356.

Sobota, A.E. 1984. Inhibition of bacterial adherence by cranberry juice: potential use for the treatment of urinary tract infections, *J. Urology*, 131(5):1013–1016.

Trevithick, J.R. and Mitton, K.P. 1999. Antioxidants and diseases of the eye, Ch. 24 in *Antioxidant Status, Diet, Nutrition, and Health*, A.M. Pappas, ed., CRC Press, Boca Raton, FL, pp. 545–565.

Tsuda, T., Horio, F., and Osawa, T. 1998. Dietary cyanidin 3-O-β-D-glucoside increases *ex vivo* oxidation resistance of serum in rats, *Lipids*, 33(6):583–587.

Tsuda, T., Ohshima, K., Kawakishi, S., and Osawa, T. 1996. Oxidation products of cyanidin 3-O-β-D-glucoside with a free radical initiator, *Lipids*, 31(12):1259–1263.

Tsuda, T., Watanabe, M., Ohshima, K., Norinobu, S., Choi, S.-W., Kawakishi, S., and Osawa, T. 1994. Antioxidative activity of the anthocyanin pigments cyanidin 3-O-β-D-glucoside and cyanidin, *J. Agric. Food Chem.*, 42(11):2407–2410.

Varma, S.D. 1986. Inhibition of aldose reductase by flavonoids: possible attentuation of diabetic complications, in *Plant Flavonoids in Biology and Medicine: Biochemical, Pharmacological, and Structure-Activity Relationships,* Alan R. Liss, Inc., pp. 343–358.

Velioglu, Y.S., Mazza, G., Gao, L., and Oomah, B.D. 1998. Antioxidant activity and total phenolics in selected fruits, vegetables, and grain products, *J. Agric. Food Chem.*, 46(10):4113–4117.

Villata, M. 1998. Cultivated blueberries: a true-blue baking ingredient, *Cereal Foods World*, 43(3):128–130.

Walker, E.B., Barney, D.P., Mickelsen, J.N., Walton, R.J. and Mickelsen Jr, R.A., 1997. Cranberry concentrate: UTI prophylaxis, *J. Fam. Prac.*, 45(2):167-168.

Wang, H., Cao, G., and Prior, R.L. 1997. Oxygen radical absorbing capacity of anthocyanins, *J. Agric. Food Chem*, 45(2):304–309.

Weiss, E.I., Lev-Dor, R., Kashamn, Y., Goldhar, J., Sharon, N., and Ofek, I. 1998. Inhibiting interspecies coaggregation of plaque bacteria with a cranberry juice constituent, *J. Am. Dent. Assoc.*, 129(12):1719–1723.

Wilson, T. and Marley, J.C. 2001. Effects of cranberry juice consumption on platelet aggregation, *FASEB J.*, 15(4):A286.

Youdim K.A., Shukitt-Hale, B., Martin, A., Bickford, P.C., and Joseph, J.A. 2000. Short-term dietary supplementation of blueberry polyphenolics: beneficial effects on aging brain performance and peripheral tissue function, *Nutr. Neuroscience,* (3):383-397.

Zafriri, D., Ofek, I., Adar, R., Pocino, M. and Sharon, N., 1989. Inhibitory activity of cranberry juice on adherence of Type 1 and Type P fimbriated *Escherichia coli* to eucaryotic cells, *Antimicrobial Agents and Chemotherapy*, 33(1):92-98.

Zaragozá, F., Iglesias, I., and Benedí, J. 1985. Estudio comparativo de los efectos antiagregantes de los antocianósidos y otros agentes, *Archiv. Farmacol. Toxicol.*, 11(3):183-188

Wine Polyphenols and Protection from Atherosclerosis and Ischemic Heart Disease: Drink To Your Health?

MICHAEL A. DUBICK

CONTENTS

INTRODUCTION

The past few years, since the passage of the Dietary Supplement Health and Education Act of 1994, have seen an explosion of interest in the development of nutriceuticals and functional foods, with the purpose of identifying foods or components of foods, such as phytochemicals, that may contribute to overall health or even treat existing disease. In this context, grapes and wine have a long history of being touted for the promotion of human health and for the treatment of disease. Recent legislation to allow wines to carry a statement citing their health benefits has the support of over 25 years of epidemiologic studies of individuals 25–84 years old which showed a U- or J-shaped relationship between incidences of myocardial infarctions, angina pectoris or coronary-related deaths, to alcohol consumption (Hegsted and Ausman, 1988; St. Leger et al., 1979; Klatsky et al., 1997; Schneider et al., 1996, Hein et al., 1997; Zakhari and Gordis, 1999; Constant, 1997). That is, at low to moderate ethanol intake, the risk of heart disease or death is lower than in abstainers, and in those who consume high levels.

This association between moderate alcohol consumption and risk of ischemic heart disease has generated what has come to be known as the "French Paradox", i.e., epidemiologic studies have observed that in southern France, mortality rates from heart disease were lower than expected despite the consumption of diets high in saturated fats and the tendency to smoke cigarettes (Drewnoswski et al., 1996; Renaud and Ruf, 1994). Aspects of the French paradox have been more completely described elsewhere (Criqui and Ringel, 1994; Renaud and De Lorgeril, 1992; Cleophas et al., 1996; Burr, 1999; Law and Wald, 1999) and are beyond the scope of this review. However, a major factor in explaining this presumed paradox is the common practice of the French consuming wine, particularly red wine, with their meals (St. Leger et al., 1979; Klatsky et al., 1997; Cleophas et al., 1996).

Although the exact mechanisms by which wine drinking could offer protection against atherosclerosis and ischemic heart disease are not fully known, a large body of literature has emerged which suggest that the actions of polyphenolic compounds found in these beverages may account for this protection (Sharp, 1993; Fero-Luzzi and Serafini, 1995; Goldberg, 1995; Hollman et al., 1999). This review will evaluate the scientific evidence for the proposed mechanisms through which polyphenols may reduce certain risk factors associated with development of cardiovascular disease (CVD).

POLYPHENOLS IN GRAPES AND WINE

Polyphenols or phenolic compounds found in the plant kingdom account for anywhere from 4000–8000 individual compounds (Fero-Luzzi and Serafini, 1995; Dresoti, 1996; King and Young, 1999; Croft, 1998; Bravo, 1998). These compounds are the secondary by-products of plant metabolism and have often evolved to help protect the plant from environmental stressors and predators. Consequently, in addition to grapes and wine, polyphenols are found in many commonly eaten fruits and vegetables, such as apples, grapefruit, and onions as well as teas (Fero-Luzzi and Serafini, 1995; Bravo, 1998).

The various polyphenolic substances in wine and grapes have been described over the years and efforts continue to better characterize their composition (Soleas and Goldberg, 1999; Hertog et al., 1993b,c; Lanningham-Foster et al., 1995). Overall, the total polyphenolic content of red wines has been estimated to be about 1200 mg/L (Soleas and Goldberg, 1999), while others have reported concentrations as high as 4000 mg/L, of which 500–900 mg/L come from the tannins (Dresoti, 2000). Part of the discrepancy reported may relate to different extraction techniques. In contrast, the polyphenolic content of white wine is about 200–300 mg/L (Frankel et al., 1995). It is also reported that the total flavonoid content of red wine can be 10- to 20-fold higher than in white wine (Soleas and Goldberg, 1999). Grape juice has about one half the flavonoid content of red wine by volume (Constant, 1997). The stilbene, resveratrol, is present mainly in grape skins, where concentrations are estimated to be 50–100 μg/g (Chun et al., 1999). Thus, resveratrol is found primarily in red wines, with concentrations around 1 mg/L (Brouillard et al., 1997). The concentrations of select flavonoids and resveratrol in wine have been summarized in a more comprehensive review (Dubick and Omaye, 2001).

It is also realized that polyphenols in aged wines may be more polymerized than those in young wines or grape juices (Bravo, 1998; Brouillard et al, 1997; Muller and Fugelsang, 1997). In addition, the amount of flavonoids in wine can be influenced by several factors, including temperature, sulfite and ethanol concentrations, the type of fermentation vessel, pH, and yeast strain, as well as the type of insecticides or herbicides used on the grapes (Muller and Fugelsang, 1997; Das et al., 1999).

POLYPHENOLS METABOLISM

In the overall evaluation of the efficacy of wine polyphenols in the prevention of atherosclerosis and heart disease, it is important to determine how well such compounds may be absorbed through the gastrointestinal tract, distributed to various target tissues, and metabolized. Unfortunately, there is limited information in humans, which has led to uncertainty that *in vivo*, these compounds could express activity of physiologic significance. Because such compounds occur as complex, highly-variable mixtures in plant materials, it is difficult to study their bioavailability and physiologic effects accurately.

Nevertheless, studies continue to investigate the absorption and metabolism of single polyphenolic compounds. In general, the absorption of polyphenols varies depending on the type of food, the chemical form of the polyphenols [(polymerized forms may be less absorbed (Muller & Fugelsang, 1997)], and their interactions with other substances in food, such as protein, ethanol, and fiber (Bravo 1998; Scalbert and Williamson, 2000). It is important to note that polyphenols from wine may be absorbed better than the same substances from fruits and vegetables, because the ethanol may enhance the breakdown of the polyphenols into smaller products that are absorbed more readily (Goldberg, 1995). The elimination of free catechin, consumed as reconstituted red wine, was increased 22% by ethanol in human subjects (Bell et al., 2000). However, since these substances rarely exist as free compounds in plasma (Manach et al., 1995, 1999), the true significance of this observation remains unknown.

Studies in experimental animals and humans indicate that some polyphenols, at least, can be absorbed (Manach et al., 1996; Manach et al., 1995; Ueno et al., 1983; de Vries et al., 1998, Clifford et al., 1996; Maiani et al., 1997; Clydsdale, 1999). As has been described elsewhere, bacteria that colonize the ileum and cecum play a significant role in the metabolism of polyphenols and are important for their absorption (Bravo, 1998; Scalbert and Williamson, 2000). In general, peak blood levels of flavonoids occur between 2 and 3 h after consumption and the elimination half-life varied between 5 and 17 h depending on the particular flavonoid or the food source (van het Hof et al., 1998; Hollman et al., 1996). However, it is important to note that some human studies of polyphenol absorption and metabolism may be misleading due to administration of pharmacologic (4 g) doses (Gugler et al., 1975). Handling of pharmacologic doses may not reflect the normal physiologic mechanisms of absorption and metabolism of dietary flavonoids that would occur with usual dietary intakes.

It should also be mentioned that evidence is accumulating that not all polyphenols need to be absorbed to be effective in reducing oxidative stress in plasma. For example, studies have shown that tannins become complexed in the gut and may not be absorbed, yet are able to reduce plasma levels of peroxides observed after consumption of a high-fat meal (Ursini et al., 1999).

ROLE OF FLAVONOIDS IN CARDIOVASCULAR DISEASE

In addition to alcohol, evidence that dietary flavonoid intake was inversely related to mortality from coronary heart disease has been supported by recent epidemiologic studies (Knekt et al., 1996; Hertog et al., 1993a, 1995, 1997; Rimm et al., 1996). In the Zutphen Elderly study, Hertog et al. (1993a, 1997) showed that after adjustment for age, weight, certain risk factors of coronary artery disease, and intake of antioxidant vitamins, the highest tertile of flavonoid intake, primarily from tea, onions, and apples, had a relative risk for heart disease of 0.32 compared with the lowest tertile, assigned a relative risk of 1.0. It should be noted that wine was not a significant contributor to flavonoid intake in these studies. At present it is not possible to distinguish from these epidemiologic observations whether the protective effect is related to polyphenols in general, consumption of certain foods high in these compounds, or other dietary components. Thus the evidence supporting a protective effect of polyphenol intake against ischemic heart disease in humans is suggestive, but still inconclusive.

As a group, the polyphenols and in particular flavonoids, have generally been recognized to possess significant antioxidant activities, such as scavenging oxygen free radicals or chelating metals (Salah et al., 1995; Rice-Evans et al., 1996; Cao et al., 1997; Sugihara et al., 1999). It is likely that various polyphenols, including flavonoids, act similar to dietary antioxidants and that collectively they may bestow protection from the development of heart disease. Other polyphenols, such as resveratrol, have been reported to have glutathione sparing mechanisms, at least *in vitro* (Burkitt and Duncan, 2000). In addition, the functional effectiveness of these compounds in protecting from ischemic or reperfusion injury has been demonstrated

TABLE 4.1
Proposed Properties of Wine or its Polyphenols to Reduce Risk of Atherosclerosis and Heart Disease

   I. General Antioxidant Activity
      Chelate transition metals
      Inhibit oxidation of LDL
      Scavenge oxygen free radicals
      Maintain plasma antioxidant vitamin levels
      ↑ Plasma uric acid (?)

  II. Effects on Plasma Lipids
      ↓ Total cholesterol
      ↑ HDL levels
      ↓ LDL levels
      ↓ Lipoprotein synthesis
      ↓ Lipoprotein [a] levels (?)

 III. Other Effects
      Anticoagulant effects including aspirin-like activity
      ↑ Nitric oxide synthesis to keep blood vessels patent
      Other anti-inflammatory activity

in various animal models (Canali et al., 2000; Sato et al., 1999, 2000; Hung et al., 2000). Table 4.1 lists the various actions suggested through which polyphenols could impact on the development of CVD.

EFFECTS ON CHOLESTEROL

Several studies in humans have suggested that the consumption of wine or ethanol was associated with lower serum total cholesterol and LDL levels, and higher HDLs (Schneider et al., 1996; Hein et al., 1997; Lavy et al., 1994). Wine or grape seed extract was also observed to be more effective than ethanol in preventing the development of atherosclerotic lesions in cholesterol-fed rabbits (Klurfeld and Kritchevsky, 1981; Yamakoshi et al., 1999). Another investigation observed that red wine consumption decreased plasma concentrations of lipoprotein [a] (Sharpe et al., 1995), an independent risk factor for atherosclerosis (Cao et al., 1998), but this observation was not supported in another study (Goldberg et al., 1996).

Studies have also supported the benefit of specific polyphenols as possibly being protective. In Japanese women, flavonoid intake, and in particular quercetin, was inversely related to plasma total and LDL cholesterol (Arai et al., 2000). Also, Xu et al (1998) reported that in hamsters fed a hypercholesterolemic diet for 16 wk, catechin supplementation was as effective as vitamin E in inhibiting plaque formation. Other recent work in hamsters suggested that the hypolipidemic activity of catechins related to their action on the absorption of dietary fat and cholesterol, rather than inhibitory effects on cholesterol or fatty acid synthesis (Chan et al., 1999).

ANTIOXIDANT PROPERTIES OF WINE

A number of *in vitro* studies have reported that wines possess intrinsic antioxidant activity. Maxwell et al. (1994) reported that red wines themselves had about 30-fold greater antioxidant activity than normal human serum, and the contribution of various polyphenols to the total antioxidant activity of red wine has been described (Rice-Evans et al., 1996). It was also observed that the total reactive antioxidant potential of red wines was 6 to 10 times higher than white wine (Campos and Lissi, 1996).

Antioxidant properties of wine have also been observed *in vivo*. For example, in nine healthy subjects who drank 300 mL of red wine, 18 and 11% increases in serum antioxidant capacity was observed after 1 h and 2 h, respectively, but less than the 22 and 29% increases seen at these times in subjects who took 1000 mg ascorbic acid (Whitehead et al., 1995). Lesser increases in serum antioxidant capacity were observed if the subjects drank white wine, or apple, grape, or orange juice. Plasma antioxidant potential also increased about 20% over baseline 2 h or 4 h after normal subjects consumed red wine or ate 1g/kg of black grapes, respectively (Durak et al., 1999), in agreement with the observation that the presence of ethanol enhances absorption of polyphenols (Ozturk et al., 1999). Others observed that drinking red wine with meals or consumption of red wine polyphenols (1 or 2 g/d) increased total plasma antioxidants by 11 to 15%, respectively, in comparison to a 7% increase by vitamin E (Maxwell et al., 1994; Nigdikar et al., 1998; Serafini et al., 1998; Struck et al., 1994). In contrast to other studies, Struck et al. (1994) observed a greater effect when the subjects drank white rather than red wine. In addition, they observed no changes in plasma vitamin E, vitamin C, or beta-carotene, but a 23% reduction from baseline in plasma retinol levels. Although these results suggest that the enhanced antioxidant potential observed after drinking wine is not due to higher concentrations of plasma antioxidant vitamins, it was reported that 73% of the increase in serum antioxidant capacity following consumption of port wine could be attributed to an increase in serum uric acid levels (Day and Stansbie, 1995), a well-recognized antioxidant (Nyyssonen et al., 1997). Caccetta et al. (2000) also observed an increase in serum uric acid levels 1 and 2 h after red wine consumption in 12 healthy males. On the other hand, Cao et al. (1998) observed an 8% increase in serum antioxidant capacity in elderly women who drank 300 ml of red wine, that could not be attributed to an increase in uric acid or vitamin C.

It should be noted that studies that did not observe an effect of red wine on plasma antioxidant status may reflect too low a consumption (Sharpe et al., 1995) or, as in a rat study, may reflect the limited effects polyphenols may exert when a well-balanced diet with more than adequate intake of micronutrients is consumed (Cestaro et al., 1996). Thus, from the above studies it would appear that polyphenols in wine and grapes demonstrate antioxidant activity, but the expression of this activity can depend on a variety of dietary and other health-related factors.

LDL OXIDATION

Citing the presumed role of oxidized LDL in the development of atherosclerosis, a large number of studies have evaluated the role of wines, grapes, or select polyphenols on inhibition of LDL oxidation. For example, Ishikawa et al. (1997) observed that catechins could inhibit LDL oxidation in a dose-dependent manner *in vitro*, and EGCG appeared to be more potent than vitamin E. Generally, studies have shown that wine polyphenols can inhibit oxidative changes of LDL, and red wine appeared more potent than white wine (Caldu et al., 1996; Rifici et al., 1999). In one study, 3.8 and 10 μM polyphenols extracted from red wine and added to LDLs *in vitro* inhibited its oxidation by 60 and 98%, respectively (Frankel et al., 1993a). Red wine also inhibited cell-mediated LDL oxidation, while white wine and ethanol were not effective (Rifici et al., 1999). Red wine, catechin, or quercetin also produced inhibitory effects on development of aortic atherosclerotic lesions, reduced the susceptibility of LDL to aggregation (Aviram and Fuhrman, 1998), and subsequent atherogenic modification of LDL in atherosclerotic apolipoprotein E-deficient mice (Hayek et al., 1997).

Polyphenols from grape extract also have the ability to inhibit oxidative changes of LDL (Yamakoshi et al., 1999; Lanningham-Foster et al., 1995). However, although red wine and grape juice could inhibit LDL oxidation, *in vitro*, LDL oxidation was only inhibited *in vivo* in those who drank wine (Miyagi et al., 1997). Using a copper chloride-induced LDL oxidative system, addition of grape extracts produced a dose-dependent prolongation of the lag time before LDL oxidation (Miyagi et al., 1997). Frankel et al. (1993a,b) further observed that the inhibition of copper-catalyzed LDL oxidation by dilutions of wine could be mimicked by equal concentrations of quercetin, but inhibition by resveratrol was only about one-half that of quercetin or epicatechin (Frankel et al., 1993b).

About eight times more white wine was required to produce a similar effect on LDL oxidation and increase HDL concentrations (Caldu et al., 1996; Rifici et al., 1999). However, if equal concentrations of wine polyphenols extracted from red or white wine were compared, inhibition of LDL oxidation was similar. Comparing dilutions of red or white wine, red grape juice, or beer, LDL oxidation was inhibited in a dose-dependent manner depending on the final polyphenol concentration (Abu-Amsha et al., 1996). These observations suggest that the reported advantage of red wine over white wine in inhibiting LDL oxidation reflects the higher concentrations of polyphenols in red than white wine, rather than red wine polyphenols being more potent than those found in white wine. However, the potencies of the various polyphenols acting alone versus potential additive or synergistic effects of the combinations found in wines have not been fully elucidated in the assay systems examined.

Despite the positive results suggesting that wine polyphenols can inhibit LDL oxidation *in vitro*, few studies have examined these effects *in vivo*. Hayek et al. (1997) fed atherosclerotic apolipoprotein-E-deficient mice red wine, quercetin, or catechin in their drinking water for 42 d. They observed that LDLs isolated from these mice were more resistant to oxidation than LDLs isolated from ethanol-fed control mice. In 15 subjects who drank 8 ml/kg of purple grape juice/d for 14 d, LDL oxidation was reduced 34.5% in comparison to controls (Stein et al., 1999). Others have also demonstrated that red wine consumption by healthy volunteers reduced the susceptibility of their LDL to oxidative damage (Lavy et al., 1994; Nigdikar et al., 1998; Seigneur et al., 1990).

The most significant *in vivo* effects of red wine consumption on inhibiting LDL oxidation was observed in 17 healthy men given 400 mL of red or white wine for 14 d. Plasma collected from red wine drinkers at the end of the study was about 20% less likely to peroxidize than plasma collected at baseline, and LDLs isolated from these subjects were more resistant to oxidation; effects independent of plasma vitamin E or ß-carotene concentrations (Fuhrman et al., 1995). However, this study has come under question because the results obtained far exceeded the clinical benefit obtained previously by dietary or pharmacologic interventions to prevent LDL oxidation (De Rijke et al., 1996). In addition, other studies have reported no significant change in LDL oxidation lag times after acute or subchronic (10–28 d) consumption of red wine despite marked increases in plasma levels of red wine polyphenols (Sharpe et al., 1995; De Rijke et al., 1996; Caccetta et al., 2000).

It should be noted that flavonoids could also accelerate LDL oxidation if they were added to minimally oxidized LDL (Otero et al., 1997). Also, some flavonoids enhanced LDL aggregation (Rankin et al., 1993), and in cholesterol-fed rabbits, resveratrol actually promoted atherosclerotic lesions in the aorta (Wilson et al., 1996). Since atherosclerotic plaque contain high concentrations of copper and iron which may catalyze LDL oxidation (Dubick et al., 1991), the net *in vivo* effect of polyphenols on LDL oxidation cannot be predicted easily. Therefore, despite the *in vitro* inhibition of LDL oxidation and the acute rise in serum antioxidant potential following consumption of red wine or grapes in particular, the limited human data do not provide strong evidence that a major *in vivo* effect of wine polyphenols is inhibition of LDL oxidation. Further work is needed to define whether these results simply reflect insufficient absorption and/or deposition of polyphenols into the target tissues, or whether different results would be obtained if these compounds were given to individuals with preexisting cardiovascular disease.

EFFECTS ON HEMOSTASIS

Wine may also have a positive influence on risk factors of CVD by inhibiting platelet aggregation and prolonging clotting times. Ethanol has long been known to exert an aspirin-like effect on clotting mechanisms (Renaud and De Lorgeril, 1992); however, the concentrations required to inhibit platelet aggregation have generally been high (Drewnoswski et al., 1996, Demrow et al., 1995). In patients with coronary artery disease, those who drank 330 mL of beer per day (~ 1 bottle/d) for 30 d showed evidence of reduced thrombogenic activity compared with patients who did not consume an alcoholic beverage (Gorinstein et al., 1987). However, this study could not determine whether these effects were due to the ethanol, polyphenols in beer, or both.

Both intravenous or intragastric administration of red wine or grape juice, but not white wine, inhibited platelet-mediated thrombus formation acutely in stenosed dog coronary arteries (Demrow et al., 1995). In addition, low concentrations (nM) of quercetin dispersed platelet thrombi that adhered to rabbit aortic endothelium *in vitro* (Gryglewski et al., 1987). It was shown that platelet aggregation was inhibited about 70% in rats fed ethanol, white wine, or red wine in their drinking water for 2 or 4 months compared with controls (Ruf et al., 1995). However, if the ethanol or wine was withheld for 18 h, platelet aggregation rebounded to greater than control levels in the ethanol- or white wine-fed groups, whereas aggregation was still inhibited in the red wine group.

Studies of the effects of wine on hemostasis in humans have not been as impressive as the animal studies. In 20 healthy men who consumed 400 mL of red or white wine for 14 d, prothrombin time increased but partial thromboplastin time decreased significantly in both groups (Lavy et al., 1994). When collagen was employed as agonist, no significant change in platelet aggregation was observed. Also, no significant effects on platelet aggregation were observed in 20 hypercholesterolemic subjects who drank red or white wine for 28 d (Struck et al., 1994). In contrast, collagen-induced platelet aggregation was lower in male volunteers who consumed 30 g of ethanol/d (about 2–3 drinks) in the form of red wine or ethanol-spiked clear fruit juice for 28 d when compared to subjects who drank dealcoholized red wine (Pellegrini et al., 1996). However, no difference in platelet aggregation was observed if ADP was employed as agonist. These data prompted the authors to conclude that their observations were due to ethanol and not to components in red wine. On the other hand, Seigneur et al. (1990) reported that ADP-induced platelet aggregation was inhibited following wine consumption. Epinephrine or arachidonic acid-induced platelet aggregation were not affected in these subjects, nor was aggregation affected in subjects who drank white wine or an ethanol solution.

In a comprehensive study by Pace-Asciak et al. (1996), 24 healthy males consumed 375 mL/d of red or white wine, or grape juice without or with added trans-resveratrol (4 mg/L) with meals for 28 d. Only white wine inhibited ADP-induced platelet aggregation, whereas red and white wine and resveratrol-supplemented grape juice inhibited thrombin-induced platelet aggregation. These authors observed that *in vitro* ADP- and thrombin-induced platelet aggregation was inhibited about 50% by grape juice without or with resveratrol, while red wine nearly abolished platelet aggregation, and white wine had no appreciable effect (Pace-Asciak et al., 1996). In a previous *in vitro* study these authors reported that platelet aggregation could be inhibited by dealcoholized red wine, quercetin, and resveratrol in a dose-dependent manner (Pace-Asciak et al., 1995).

Most recently, in 10 healthy subjects drinking 5–7.5 ml/kg/d of grape juice for 1 wk, whole blood platelet aggregation was reduced 77% when collagen was used as agonist (Keevil et al., 2000). However, consuming 1 g/d of quercetin supplements did not affect collagen-induced platelet aggregation in healthy men and women (Conquer et al., 1998). Again, these data suggest that the *in vitro* effects of wine and its polyphenols on platelet aggregation are more pronounced than the *in vivo* effects in humans. Nevertheless, it remains to be determined whether these effects on platelet aggregation are of clinical importance and can translate into reduced risk from thrombi formation and the risk of a coronary event.

NITRIC OXIDE-RELATED EFFECTS

Nitric oxide (NO) is a major mediator of vascular relaxation that also inhibits platelet adherence to endothelium. Evidence suggests that wine polyphenols may modulate the production of nitric oxide, since wines, grape juice, and extracts from grape skins relaxed precontracted rat aortic rings (Andriambeloson et al., 1997; 1998). In addition, quercetin could reproduce the effects of wine and grape fractions, while resveratrol, catechin, and malvidin could not (Fitzpatrick et al., 1995; Keaney, 1995). In human subjects who consumed grape juice for 14 d, endothelial-dependent vasodilation was about three-fold higher than in controls (Stein et al., 1999). Taken together these results are interesting, but most of the work has been *in vitro* and further studies are needed to define the role of polyphenols in inducing vasodilation, particularly after *in vivo* consumption, and it remains unknown whether such effects could translate into reduced risk for CVD.

In addition, both red and white wine contain significant amounts of salicylic acid and its dihydroxybenzoic acid metabolites, which also have vasodilator and anti-inflammatory activities (Muller and Fugelsang, 1994). The concentrations of these compounds in wine range from 11–28.5 mg/L, with the concentrations being higher in red than white wine. Again, it remains to be determined whether these compounds are absorbed after drinking wine or whether plasma concentrations attained would have the physiologic effect observed *in vitro*.

POTENTIAL ADVERSE EFFECTS OF POLYPHENOLS

Before consumption of polyphenols as a supplement or in wine can be recommended as part of a dietary regimen to reduce risk factors associated with CVD, it is important to review any evidence related to adverse effects. Generally, consumption of polyphenols through a variety of foods is not likely to produce adverse effects, because of the diversity and varying quantities of polyphenols in plant sources. However, evidence suggests that flavonoids may cross the placenta and become concentrated in the developing fetus and perhaps increase the risk of developing infantile leukemia. Therefore, consumption of large doses of polyphenols probably should be avoided during pregnancy, but this area has received little attention. In addition, chronic pharmacologic doses have been reported to produce adverse effects. For example, doses of 1–1.5 g/d of cianidanol, a flavonoid drug, produced renal failure, hepatitis, fever, hemolytic anemia, thrombocytopenia, and skin disorders (Jaeger et al., 1988). Also, high doses of flavonoids can induce mutagenic effects, produce free radicals, and inhibit enzymes involved in hormone metabolism (Skibola and Smith,

2000). Considering that polyphenols are redox chemicals, these activities may also be due to the concomitant production of hydrogen peroxide by phenolics during their autoxidation in a process that is dependent on divalent metal ions, similar to the metal-induced pro-oxidant effect of vitamin C (Stadler et al., 1995; Arizza and Pueyo, 1991).

The potential adverse effects of some individual polyphenols have also been studied. For example, in subjects consuming about 1 g/d EGCG supplements, approximately the dose found in people who drink >10 cups of green tea/d, some stomach discomfort was noted that resolved if the tablets were taken after a meal. Some transient sleeplessness was also reported, but could be due to caffeine contamination of the extract. The LD50 in rats is reported to be 5g/kg in males and 3.1 g/kg in females, suggesting that EGCG has relatively low acute toxicity, but may express teratogenicity at concentrations potentially achievable with daily consumption. In addition, sensitivity to EGCG reportedly has induced asthma in workers at a green tea factory (Clydesdale, 1999). Thus, concerns of allergic reactions, much like those reported with herbal teas, may need to be considered in susceptible individuals.

Quercetin also appears to be relatively nontoxic with an LD50 in mice over 100 mg/kg. In a phase I clinical study, nephrotoxicity was not observed until a cumulative dose of 1700 mg/m^2 was achieved (Clydesdale, 1999). No evidence of carcinogenicity or teratogenicity with quercetin has been reported, even when fed at dietary levels as high as 10% (Clydesdale, 1999).

A study in Portugal observed a dose-dependent relationship between red wine consumption and incidence of gastric cancer (Falcao et al., 1994). However, it is not known if this observation is related to the ethanol, red wine polyphenols, or their interaction with other risk factors. Free radicals have been identified in red wines and their originating grape source, but have not been detected in white wine (Troup et al., 1994). Flavonoids can also express pro-oxidant effects in the presence of copper (Cao et al., 1997) or NO (Ohshima et al., 1998), and Halliwell (1993) reported that plant phenolics may show an oxidant effect against proteins and DNA. Conditions where phenolic compounds act as pro-oxidants have been described (Decker, 1997; Laughton et al., 1989).

CONCLUSIONS

The results from the studies summarized here indicate that the polyphenols present in grapes and wine, among other foods, possess antioxidant activity, and have the potential to modify plasma cholesterol and lipoprotein concentrations, inhibit LDL oxidation, reduce platelet aggregation, and have vasorelaxant effects, i.e., modify certain risk factors associated with the development of atherosclerosis or ischemic heart disease in susceptible individuals. Although these effects have been well demonstrated *in vitro*, in most cases the *in vivo* results have been less convincing. These differences may simply reflect low rates of absorption of the active compounds, differences in methodologies employed by the various investigators to detect these effects, the antioxidant status of the subjects, or other factors. In most instances, studies in humans have employed healthy subjects, and it is unknown whether potential benefit would be observed in individuals with varying stages of CVD. It is also realized that wines are complex mixtures and their polyphenol content varies. Even in cases where specific polyphenols are studied, the results have not always been consistent. In studies where red wine is reportedly superior to white wine, the differences most likely merely reflect the higher polyphenol concentrations in red than white wine, rather than differences in the potencies of polyphenols that may be found in both wines.

To date, in studies of increases in antioxidant potential in plasma after wine consumption, the changes have been transient and disappear a few hours after drinking. In the platelet aggregation studies, results have been reported for only 1 or 2 time points after drinking, and aggregation may change in response to one agonist, but not another, making the overall physiologic significance difficult to interpret. Also, it remains to be established whether these results would be sustained after continued consumption, or whether adaptation to these effects would occur.

As mentioned above, polyphenols are also present in a number of common fruits and vegetables (Fero-Luzzi and Serafini, 1995; Bravo, 1998; Hertog et al., 1995). Numerous studies have touted the potential health benefits of consuming diets rich in fruits and vegetables, particularly with regard to cancer prevention. Since it is unclear which of the polyphenols offer the most health-promoting advantage, it would seem premature and inappropriate to recommend drinking wine specifically as a major source of polyphenols in an attempt to raise an individual's plasma antioxidant status as a means to reduce risk of CVD. Even in the case where wine may contain a particular polyphenol such as resveratrol, not generally found in other common foods, the evidence that it possesses any specific protective effects against CVD *in vivo*, is insufficient. On the other hand, there is sufficient information to suggest that for adults, consuming 1 to 2 glasses of wine with meals would not be harmful, and may be beneficial (Meister, 1999). In considering any recommendation for wine, one must always keep in mind the well-known adverse health effects and consequences of chronic ethanol abuse or the effects and risks of acute inebriation (Zakhari and Gordis, 1999; Dufour and Fe Caces, 1993; Rubin, 1993).

From the forgoing discussion, it would appear that despite a wealth of *in vitro* studies citing the potential efficacy of grape and wine polyphenols against known risk factors of CVD, much research is required to confirm benefit *in vivo*. Many of these studies employed high doses of flavonoids, and additional studies could address repeated administration of lower doses and/or identify definitively which polyphenols in wine and grapes may be most important for human health. It is also unknown whether individual polyphenols can act in an additive or synergistic fashion *in vivo*. Since some polyphenols are mutually exclusive in nature (Rice-Evans et al., 1996), it is not known whether packaging them together in a supplement would be potentially harmful.

In summary, the available data to date indicate that grape and wine polyphenols, as well as the complex beverage, possess biologic activity that may potentially modify certain risk factors associated with atherogenesis and CVD. Indeed, these polyphenols have the potential to be an important source of nonnutritive antioxidants in the diet (Prior and Cao, 1999). Epidemiologic studies showing an inverse relationship between flavonoid intake and incidence of CVD suggest that individuals 35 to 65 or 70 years of age may benefit most from these compounds. Unfortunately, the data are slim to suggest convincingly that these substances may offer long-term protection from these diseases. There is also no definitive evidence that wine consumption would be beneficial in trying to overcome the adverse health consequences of smoking or other factors associated with an unhealthy lifestyle. Since the best polyphenols to promote health are not known, it also seems premature to recommend dietary supplements containing individual compounds or complexes at this time, particularly since standard Western diets may contain up to 1 g/d of polyphenols (Scalbert and Williamson, 2000). Also, in view of potential adverse effects of various polyphenols, caution is advised against long term intake of gram doses of individual compounds above normal daily dietary levels, and certainly in pregnant women (Skibola and Smith, 2000). As usual, variety, moderation, and balance should remain the best recommendation when considering the addition of wine or various polyphenols to the diet.

ACKNOWLEDGMENT

The author thanks Amber Large for assistance in the preparation of the manuscript.

REFERENCES

Abu-Amsha, R., Croft, K.D., Puddey, I.B., Proudfoot, J.M., and Beilin, L.J. 1996. Phenolic content of various beverages determines the extent of inhibition of human serum and low-density lipoprotein oxidation *in vitro:* identification and mechanism of action of some cinnamic acid derivatives from red wine, *Clin. Sci.*, 91:449–458.

Andriambeloson, E., Kleschyov, A.L., Muller, B., Beretz, A., Stoclet, J.C., and Andriantsitohania, R. 1997. Nitric oxide production and endothelium-dependent vasorelaxation induced by wine polyphenols in rat aorta, *Brit. J. Pharmacol.*, 120:1053–1058.

Andriambeloson, E., Magnier, C., Haan-Archipoff, G., Lobstein, A., Anton, R., Beretz, A., Stoclet, J.C., and Andriantsitohaina, R. 1998. Natural dietary polyphenolic compounds cause endothelium-dependent vasorelaxation in rat thoracic aorta, *J. Nutr.*, 128:2324–2333.

Arai, Y., Wantanabe, S., Kimira, M., Shimoi, K., Mochizuki, R., and Kinae, N. 2000. Dietary intake of flavonoids, flavones, and isoflavones by Japanese women and the inverse correlation between Quercetin intake and plasma LDL cholesterol concentration, *J. Nutr.*, 130:2243–2250.

Arizza, R.R. and Pueyo, C. 1991. The involvement of reactive oxygen species in the direct acting mutagencity of wine, *Mutat. Res.*, 251:115–121.

Aviram, M. and Fuhrman, B. 1998. Polyphenolic flavonoids inhibit macrophage-mediated oxidation of LDL and attenuate atherogenesis, *Atherosclerosis*, 13:S45–S50.

Bell, J.R.C., Donovan, J.L., Wong, R., Waterhouse, A.L., German, J.B., Walzem, R.L., and Kasim-Karakas, S.E. 2000. Catechin in human plasma after ingestion of a single serving of reconstituted red wine, *Am. J. Clin. Nutr.,* 71:103–108.

Bravo, L. 1998. Polyphenols: chemistry, dietary sources, metabolism, and nutritional significance, *Nutr. Rev.,* 56:317–333.

Brouillard, R., George, F., and Fougerousse, A. 1997. Polyphenols produced during red wine ageing, *Biofactors,* 6:403–410.

Burkitt, M.J. and Duncan, J. 2000. Effects of trans-resveratrol on copper-dependent hydroxyl-radical formation and DNA damage: evidence for hydroxyl-radical scavenging and a novel, glutathione-sparing mechanism of action, *Arch. Biochem. Biophys.,* 381:253–263.

Burr, M.L. 1999. Explaining the French paradox, *J. R. Soc. Health,* 115:217–219.

Caccetta, R.A.-A., Croft, K.D., Beilin, L.J., and Puddey, I.B. 2000. Ingestion of red wine significantly increases plasma phenolic acid concentrations but does not acutely affect *ex vivo* lipoprotein oxidizability, *Am. J. Clin. Nutr.,* 71:67–74.

Caldu, P., Hurtado, I., and Fiol, C. 1996. White wine reduces the susceptibility of low-density lipoprotein to oxidation, *Am. J. Clin. Nutr.,* 63:403.

Campos, A.M. and Lissi, E.A. 1996. Total antioxidant potential of Chilean wines, *Nutr. Res.,* 16:385–389.

Canali, R., Vignolini, F., Nobili, F., and Mengheri, E. 2000. Reduction of oxidative stress and cytokine-induced neutrophil chemoattractant (CINC) expression by red wine polyphenols in zinc deficiency intestinal damage of rat, *Free Rad. Biol. Med.,* 28:1661–1670.

Cao, G., Russell, R.M., Lischner, N., and Prior, R.L. 1998. Serum antioxidant capacity is increased by consumption of strawberries, spinach, red wine or vitamin C in elderly women, *J. Nutr.,* 128:2383–2390.

Cao, G., Sofic, E., and Prior, C. 1997. Antioxidant and pro-oxidant behavior of flavonoids: structure–activity relationships, *Free Rad. Biol. Med.,* 22:749–760.

Cestaro, B., Simonetti, P., Cervato, G., Brusamolino, A., Gatti, P., and Testolin, G. 1996. Red wine effects on peroxidation indexes of rat plasma and erythrocytes, in *J. Food. Sci. Nutr.,* 47:181–189.

Chan, P.T., Fong, W.P., Cheung, Y.L., Huang, Y., Ho, W.K., and Chen, Z.Y. 1999. Jasmine green tea epicatechins are hypolipidemic in hamsters (*MesocritcetusAuratus*) fed a high-fat diet, *J. Nutr.,* 129:1094–1101.

Chun, Y.J., Kim, M.Y., and Guengerich, F.P. 1999. Resveratrol is a selective human cytochrome P450 1A1 inhibitor, *Biochem. Biophys. Res. Commun.,* 262:20–25.

Cleophas, T.J., Tuinenberg, E., Van Der Meulen, J., and Zwinderman, K.H. 1996. Wine consumption and other dietary variables in males under 60 before and after acute myocardial infarction, *Angiology,* 47:789–796.

Clifford, A.J., Ebeler, J.D., Bills, N.D., Hinrichs, S.H., Teissedre, P.L., Waterhouse, A.L. 1996. Delayed tumor onset in transgenic mice fed an amino acid-based diet supplement with red wine solids, *Am. J. Clin. Nutr.,* 64:748–756.

Clydesdale, F.M. (ed). 1999. Epigallocatechin and epigallocatechin-3-gallate, *Crit. Rev. Food Sci. Nutr.,* 39:215–226.

Conquer, J.A., Miaani, G., Azzini, E., Raguzzini, A., Holub, B.J. 1998. Supplementation with Quercetin markedly increases plasma Quercetin concentration without effect on selected risk factors for heart disease in healthy subjects, *J. Nutr.,* 128:593–597.

Constant, J. 1997. Alcohol, ischemic heart disease, and the French Paradox, *Clin. Cardiol.,* 20:420–424.

Criqui, M.H. and Ringel, B.L. 1994. Does diet or alcohol explain the French Paradox, *Lancet*, 34:1719–1723.

Croft, K.D. 1998. The chemistry and biological effects of flavonoids and phenolic acids, *Ann. N.Y. Acad. Sci.*, 854:435–442.

Das, D.K., Sato, M., Ray, P.S., Maulik, G., Engelman, R.M., and Bertelli, A.A. 1999. Cardioprotection of red wine: role of polyphenolic antioxidants, *Drugs Exp. Clin. Res.*, 25:115–120.

Day, A. and Stansbie, D. 1995. Cardioprotective effect of red wine may be by urate, *Clin. Chem.*, 41:1319–1320.

Decker, E.A. 1997. Phenolics: pro-oxidants or antioxidants, *Nutr. Rev.*, 55:396–398.

Demrow, H.S., Slane, P.R., and Folts, J.D. 1995. Administration of wine and grape juice inhibits *in vivo* patients' activity and thrombosis in stenosed canine arteries, *Circulation*, 91:1182–1188.

De Rijke, Y.B., Demacker, P., Assen, N.A., Sloots, L.M., Katan, M.B., and Stalenhoef, A.F.H. 1996. Red wine consumption does not effect oxidizabilitiy of low-density lipoproteins in volunteers, *Am. J. Clin. Nutr.*, 63:329–334.

de Vries, J.H.M., Hollman, P.C.H., Meyboom, S., Buysman, M.N.C.P., Zock, P.L., van Staveren, W.A., and Katan, M.B. 1998. Plasma concentrations and urinary excretion of the antioxidant flavonols Quercetin and Kaempferol as biomarkers for dietary intake, *Am. J. Clin. Nutr.*, 68:60–65.

Dresoti, I.E. 1996. Bioactive ingredients: antioxidants and polyphenols in tea, *Nut. Rev.*, 54:S51–S58.

Dresoti, I.E. 2000. Antioxidant polyphenols in tea, cocoa, and wine, *Nutr.*, 16:692–694.

Drewnoswski, A., Henderson, S.A., and Shore, A.B., et al. 1996. Diet quality and dietary diversity in France: implications for the French Paradox, *J. Am. Diet Assoc.*, 96:663–669.

Dubick, M.A., Hunter, G.C., and Keen, C.L. 1991. Trace element and mineral concentrations in serum and aorta from patients with abdominal aneurysmal or occlusive disease, *J. Trace Elem. Exp. Med.*, 4:173–182.

Dubick, M.A. and Omaye, S.T. 2001. Modification of atherogensis and heart disease by grape wine and tea polyphenols, in *Handbook of Nutraceuticals and Functional Foods* (Wildman REC, ed.), CRC Press, Boca Raton, FL, pp. 235–260.

Dufour, M.C. and Fe Caces, M. 1993. Epidemiology of the medical consequences of alcohol, *Alcohol Health Res. Wld.*, 17:265–271.

Durak, I., Koseoglu, M.H., Kacmaz, M., Buyukkocak, S., Cimen, B., and Ozturk, H.S. 1999. Black grape enhances plasma antioxidant potential, *Nutr. Res.*, 19:973–977.

Falcao, J.M., Dias, J.A., Miranda, A.C., Leitao, C.N., Lacerda, M.M., and da Motta, L.C. 1994. Red wine consumption and gastric cancer in Portugal: a case-control study, *Eur. J. Cancer Prev.*, 3:269–276.

Fero-Luzzi, A. and Serafini, M. 1995. Polyphenols in our diet: do they matter, *Nutrition*, 11:399–400.

Fitzpatrick, D.F., Hirschfield, S.L., and Coffey, R.G. 1995. Endothelium-dependent vasorelaxing activity of wine and other grape products, *Am. J. Physiol.*, 265:H774–H778.

Frankel, E.N., Kanner, J., German, J.B., Parks, E., and Kinsella, J.E. 1993a. Inhibition of oxidation of human low-density lipoprotein by phenolic substances in red wine, *Lancet*, 341:454–457.

Frankel, E.N., Waterhouse, A.L., and Kinsella, J.E. 1993b. Inhibition of human LDL oxidation by resveratrol, *Lancet*, 341:1103–1104.

Frankel, E.N., Waterhouse, A.L., and Teissedre, P.L. 1995. Principal phenolic phytochemicals in selected California wines and their antioxidant activity in inhibiting oxidation of human low-density lipoproteins, *J. Agric. Food Chem.*, 43:890–894.

Fuhrman, B., Lavy, A., and Aviram, M. 1995. Consumption of red wine with meals reduces the susceptibility of human plasma and low-density lipoprotein to lipid peroxidation, *Am. J. Clin. Nutr.*, 61:549–554.

Goldberg, D.M. 1995. Does wine work? *Clin. Chem.*, 41:14–16.

Goldberg, D.M., Garovic-Kocic, V., Diamandis, E.P., and Pace-Asciak, C.R. 1996. Wine: does color count? *Clin. Chim. Acta,* 246:183–193.

Gorinstein, S., Zemser, M., Lichman, I., Berebi, A., Kleipfish, A., Libman, I., Trakhtenberg, S., and Caspi, A. 1987. Moderate beer consumption and the blood coagulation in patients with coronary artery disease, *J. Intern. Med.*, 241:47–51.

Gryglewski, R.J., Korbut, R., Robak, J., and Swies, J. 1987. On the mechanism of antithrombotic action of flavonoids, *Biochem. Pharm.*, 36:317–322.

Gugler, R., Leschik, M., and Dengler, H.L. 1975. Disposition of quercetin in man after single oral and intravenous doses, *Eur. J. Clin. Pharmacol.*, 9:229–234.

Halliwell, B. 1993. Antioxidants in wine, *Lancet*, 341:1538.

Hayek, T., Fuhrman, B., Vaya, J., Roenblat, M., Belinky, P., Coleman, R., Elis, A., and Aviram, M. 1997. Reduced progression of atherosclerosis in apolipoprotein E-deficient mice following consumption of red wine, or its polyphenols quercetin or catechin, is associated with reduced susceptibility of LDL to oxidation and aggregation, *Arterioscler. Thromb. Vasc. Biol.*, 17:2744–2752.

Hegsted, M.D. and Ausman, L.M. 1988. Diet, alcohol and coronary heart disease in Men, *J. Nutr.*, 118:1184–1189.

Hein, H.O., Suadicani, P., Gyntelbery, F. 1997. Alcohol consumption, S-LDL-cholesterol and risk of heart disease. 6-year follow-up in the Copenhagen male study, *Ugeskritt Laeger,* 159:4110–4416.

Hertog, M.G.L., Feskens, E.J.M., Hollman, P.C.H., Katan, M.B., and Kromhout, D. 1993a. Dietary antioxidant flavonoids and risk of coronary heart disease: the Zutphen elderly study, *Lancet*, 342:1007–1011.

Hertog, M.G.L., Feskens, E.J.M., Kromhout, D. 1997. Antioxidant flavonols and coronary heart disease risk, *Lancet*, 349:699.

Hertog, M.G., Hollman, P.C., Katan, M.B., and Kromhout, D. 1993b. Intake of potentially anticarcinogenic flavonoids and their determinants in adults in the Netherlands, *Nutr. Cancer*, 20:21–29.

Hertog, M.G.L., Hollman, P.C.H., and van de Purtte, B. 1993c. Content of potentially anticarcinogenic flavonoids of tea infusions, wines, and fruit juices, *J. Agric. Food Chem.*, 41:1242–1246.

Hertog, M.G.L., Kromhout, K., Aravanis, C., Blackburn, H., Buzina, R., Fidanza, F., Giampaolis, S., Jansen, A., Menotti, A.L., Nedeljkovic, S., Pekkarinen, M., Simic, B.O., Toshima, H., Feskens, E.J.M., Hollman, P.C.H., and Katan, M.B. 1995. Flavonoid intake and long-term risk of coronary heart disease and cancer in the seven countries study, *Arch. Int. Med.,* 155:381–386.

Hollman, P.C.H., Feskens, E.M.J., and Katan, M.B. 1999. Tea flavonols in cardiovascular disease and cancer epidemiology, *Proc. Soc. Exp. Biol. Med.*, 220:198–202.

Hollman, P.C.H., Van der Gaag, M., and Mengelers, M.J.B. 1996. Absorption and disposition kinetics of the dietary antioxidant quercetin in man, *Free Rad. Biol. Med.*, 21:703–707.

Hung, L.-M., Chen, J.-K., Huang, S.-S., Lee, R.-S., Su, M.-J. 2000. Cardioprotective effect of resveratrol, a natural antioxidant derived from grapes, *Cardiovascular Res.*, 47:549-555.

Ishikawa, T., Suzukawa, M., Ito, T., Yoshida, H., Ayaori, M., Nishiwaki, M., Yonemura, A., Hara, Y., and Nakamura, H. 1997. Effect of tea flavonoid supplementation on the susceptibility of low-density lipoprotein to oxidative modification, *Am. J. Clin. Nutr.*, 66:261–266.

Jaeger, A., Walti, M. and Neftel, K. 1988. Side effects of flavonoids in medical practice, *Prog. Clin. Biol. Res.*, 280:379–394.

Keaney, J.F., Jr. 1995. Atherosclerosis, oxidative stress, and antioxidant protection in endothelium-derived relaxing factor action, *Prog. Cardiovasc. Dis.*, 38:129–154.

Keevil, J.G., Osman, H.E., Reed, J.D., Folts, J.D. 2000. Grape juice, but not orange juice or grapefruit juice, inhibits human platelet aggregation, *J. Nutr.*, 130:53–56.

King, A. and Young, G. 1999. Characteristics and occurrences of phenolic phytochemicals, *J. Am. Diet Assoc.*, 99:213–218.

Klatsky, A.L., Armstrong, M.A., and Friedman, G.D. 1997. Red wine, white wine, liquor, beer and risk for coronary artery disease hospitalization, *Am. J. Cardiol.*, 80:416–420.

Klurfeld, D.M. and Kritchevsky, D. 1981. Differential effect of alcoholic beverages on experimental atherosclerosis in rabbits, *Exp. Mol. Pathol.*, 34:62–71.

Knekt, P., Jarvinen, R., Reunanen, A. and Maatela, J. 1996. Flavonoid intake and coronary mortality in Finland: a cohort story, *BMJ*, 312:478–481.

Lanningham-Foster, L., Chen, C., Chance, D.S., and Loo, G. 1995. Grape extract inhibits lipid peroxidation of human low density lipoprotein, *Biol. Phar. Bull.*, 18:1347–1351.

Laughton, M.L., Halliwell, B., Evans, P., Robin, J., and Hoult, S. 1989. Antioxidant and prooxidant actions of the plant phenolics quercetin, gossypol and myricetin. Effects on lipid peroxidation, hydroxyl radical generation and bleomycin–dependent damage to DNA, *Biochem. Pharmacol.*, 38:2859–2865.

Lavy, A., Fuhrman, B., Markel, A., Dankner, G., Amotz, A.B., Presser, D., and Aviram, M. 1994. Effect of dietary supplementation of red or white wine on human blood chemistry, hematology and coagulation: favorable effect of red wine on plasma high-density lipoprotein, *Ann. Nutr. Metab.*, 38:287–294.

Law, M. and Wald, N. 1999. Why heart disease mortality is low in France: the time lag explanation, *BMJ*, 318:1471–1480.

Maiani, G., Serafini, M., Salucci, M., Axxini, E., Ferro-Luzzi, A. 1997. Application of new high-performance liquid chromatographic method for measuring selected polyphenols in human plasma, *J. Chromat. B. Biomed. Sci. Appl.*, 692:311–317.

Manach, C., Morand, C., Texier, O., Agullo, G.A., Demigne, C., and Ramsey, C. 1996. Bioavailability, metabolism, and physiological impact of 4-oxo-flavonoids, *Nutr. Res.*, 16:517–544.

Manach, C., Morand, C., Texier, O., Favier, M.L., Agullo, G., Demigne, C., Regerat, F., and Remesy, C. 1995. Quercetin metabolites in plasma of rats fed diets containing rutin or quercetin, *J. Nutr.*, 125:1911–1922.

Manach, C., Texier, O., Morand, C., Crespy, V., Francoise R., Demigne, C., and Remesy, C. 1999. Comparison of the bioavailability of quercetin and catechin in rats, *Free Rad. Biol. Med.*, 27:1259–1266.

Maxwell, S., Cruickshank, A., and Thorpe, G. 1994. Red wine and antioxidant activity in serum, *Lancet*, 344:193–194.

Meister, K. 1999. Moderate Alcohol Consumption and Health, American Council on Science and Health Report, New York.

Miyagi, Y., Miwa, K., and Inoue, H. 1997. Inhibition of human low-density lipoprotein oxidation by flavonoids in red wine and grape juice, *Am. J. Cardiol.*, 80:1627–1631.

Muller, C.J. and Fugelsang, K.C. 1994. Take two glasses of wine and see me in the morning, *Lancet*, 334:1428–1429.

Muller, C.J. and Fugelsang, K.C. 1997. Red wine but not white: the importance of fully characterizing wines used in health studies, *Am. J. Clin. Nutr.*, 66:447–51.

Nigdikar, S.V., Williams, N.R., Griffin, B.A., and Howard, A.N. 1998. Consumption of red wine polyphenols reduces the susceptibility of low-density lipoproteins to oxidation *in vivo, Am. J. Clin. Nutr.*, 68:258–265.

Nyyssonen, K., Porkkala-Sarataho, E., and Salonen, J.T. 1997. Ascorbate and urate are the oxidation in Finnish men, *Atherosclerosis*, 130:223–233.

Ohshima, H., Yoshie, Y., Auriol, S., and Gilibert, I. 1998. Antioxidant and pro-oxidant actions of flavonoids: effects on DNA damage induced by nitric oxide, peroxynitrite and nitroxyl anion, *Free Rad. Biol. Med.*, 25:1057–1065.

Otero, P., Viana, M., Herrera, E., and Bonet, B. 1997. Antioxidant and prooxidant effects of ascorbic acid, dehydroascorbic acid and flavonoids on LDL submitted to different degrees of oxidation, *Free Rad. Res.*, 27:619–626.

Ozturk, H.S., Kacmaz, M., Cimen, M.Y.B., and Durak, I. 1999. Red wine and black grape strengthen blood antioxidant potential, *Nutr.*, 15:954.

Pace-Asciak, C.R., Hahn, S., Diamandis, E.P., Soleas, G., and Goldberg, D.M. 1995. The red wine phenolics trans-resveratrol and quercetin block human platelet aggregation and eicosanoid synthesis: implications for protection against coronary heart disease, *Clin. Chim. Acta,* 235:207–219.

Pace-Asciak, C.R., Rounova, O., Hahn, S.E., Diamandis, E.P., and Goldberg, D.M. 1996. Wines and grape juices as modulators of platelet aggregation in healthy human subjects, *Clin. Chim. Acta*, 246:163–182.

Pellegrini, N., Pareti, F.I., Stabile, F., Brusamolino, A., Simonetti, P. 1996. Effects of moderate consumption of red wine on platelet aggregation and haemostatic variables in healthy volunteers, *Eur. J. Clin. Nutr.*, 50:209–213.

Prior, R.L., Cao, G. 1999. Antioxdiant capacity and polyphenolic components of teas: implications for altering *in vivo* antioxidants status, *Proc. Soc. Exp. Biol. Med.*, 220:255–261.

Rankin, S.M., de Whalley, C.V., Hoult, J.R., Jessup, W., Wilkins, G.M., Collard, J., and Leake, D.S. 1993. The modification of low density lipoprotein by the flavonoids myricetin and gossypetin, *Biochem. Pharm.*, 45:67–75.

Renaud, S., De Lorgeril, M. 1992. Wine, alcohol, platelets, and the French Paradox for coronary heart disease, *Lancet*, 339:1523–1526.

Renaud, S. and Ruf, J.C. 1994. The French Paradox: vegetables or wine, *Circulation*, 90:3118.

Rice-Evans, C.A., Miller, N.J., and Paganaga, G. 1996. Structure-antioxidant activity relationships of flavonoids and phenolic acids, *Free Rad. Biol. Med.*, 20:933–956.

Rifici, V.A., Stephan, E.M., Schneider, S.H., and Khachadurian, A.K. 1999. Red wine inhibits the cell-mediated oxidation of LDL and HDL, *J. Am. Coll. Nutr.*, 18:137–143.

Rimm, E.B., Katan, M.B., Ascherio, A., Stampfer, M.J., and Willett, W.C. 1996. Relation between intake of flavonoids and risk of coronary heart disease in male health professionals, *Ann. Int. Med.*, 125:384–389.

Rubin, E. 1993. The chemical pathogenesis of alcohol-induced tissue injury, *Alcohol Health Res. Wld.*, 17:272–278.

Ruf, J.C., Berger, J.L., and Renaud, S. 1995. Platelet rebound effect of alcohol withdrawal and wine drinking in rats, *Arterioscler. Thromb. Vasc. Biol.,* 15:140–144.

St. Leger, A.S., Cochrane, A.L., and Moore, F. 1979. Factors associated with cardiac mortality in developed countries with particular reference to the consumption of wine, *Lancet,* 1017–1020.

Salah, N., Miller, N.J., Paganaga, G., Tijburg, L., Bolwell, G.P., and Rice-Evans, C. 1995. Polyphenolic flavanols as scavengers of aqueous phase radicals and as chain-breaking antioxidants, *Arch. Biochem. Biophys.,* 322:339–349.

Sato, M., Maulik, G., Ray, P.S., Bagchi, D., Das, D.K. 1999. Cardioprotective effects of grape seed proanthocyanidin against ischemic reperfusion injury, *J. Mol. Cell. Cardiol.,* 31:1289–1297.

Sato, M., Ray, P.S., Maulik, G., Maulik, N., Engelman, R.M., Bertelli, A.E., Bertelli, A., and Das, D.K. 2000. Myocardial protection with red wine extract, *J. Cardio. Pharm.,* 35:263–268.

Scalbert, A. and Williamson, G. 2000. Dietary intake and bioavailability of polyphenols, *J. Nutr.,* 130:2073S–2085S.

Schneider, J., Kaffarnik, H., and Steinmetz, A. 1996. Alcohol, lipid metabolism and coronary heart disease, *Herz,* 21:217–226.

Seigneur, M., Bonnet, J., Dorian, B., Benchimol, D., Drouiller, F., Gouverneur, G., Larrue, J., Crockett, R., Boisseau, M.R., Ribereau-Gayon, P., and Bricaud, H. 1990. Effect of the consumption of alcohol, white wine, and red wine on platelet function and serum lipid, *J. Appl. Cardiol.,* 5:215–222.

Serafini, M., Maiani, G., and Ferro-Luzzi, A. 1998. Alcohol-free red wine enhances plasma antioxidant capacity in humans, *J. Nutr.,* 128:1003–1007.

Sharp, D. 1993. When wine is red, *Lancet,* 341:27–28.

Sharpe, P.C., McGrath, L.T., McClean, E., Young, I.S., and Archbold, G.P.R. 1995. Effect of red wine consumption on lipoprotein (a) and other risk factors for atherosclerosis, *Q. J. Med.,* 88:101–108.

Skibola, C.F. and Smith, M.T. 2000. Potential health impacts of excessive flavonoid intake, *Free Rad. Biol. Med.,* 29:375–383.

Soleas, G.J. and Goldberg, D.M. 1999. Analysis of antioxidant wine polyphenols by gas chromatography-mass spectrometry, *Mthd. Enzymol.,* 299:137–151.

Stadler, R.H., Markovic, J., and Turesky, R.J. 1995. *In vitro* anti- and pro-oxidative effects of natural polyphenols, *Biol. Trace Element Res.,* 47:299–305.

Stein, J.H., Keevil, J.G., Wiebe, D.A., Aeschilmann, S., and Folts, J.D. 1999. Purple grape juice improves endothelial function and reduces the susceptibility of LDL cholesterol to oxidation in patients with coronary artery disease, *Circ.,* 100:1050–1055.

Struck, M., Watkins, T., Tomeo, A., Halley, J., and Bierenbaum, M. 1994. Effects of red and white wine on serum lipids, platelet aggregation, oxidation products and antioxidants: a preliminary report, *Nutr. Rev.,* 14:1811–1819.

Sugihara, N., Arakawa, T., Ohnishi, M., and Furuno, K. 1999. Anti- and pro-oxidative effects of flavonoids on metal-induced lipid hydroperoxide-dependent lipid peroxidation in cultured hepatocytes loaded with α-linolenic acid, *Free Rad. Biol. Med.,* 27:1313–1323.

Troup, C.J., Hutton, D.R., Hewitt, D.G., and Hunter, C.R. 1994. Free radicals in red wine, but not in white? *Free Rad. Res.,* 20:63–68.

Ueno, I., Nakano, N., and Hirono, I. 1983. Metabolic fate of [14C] quercetin in the ACI rat, *Jpn. J. Exp. Med.,* 53:41–50.

Ursini, F., Tubaro, F., Rong, J., and Sevanian, A. 1999. Optimization of nutrition: polyphenols and vascular protection, *Nutr. Rev.*, 57:241–249.

van het Hof, L.H., Kivits, G.A.A., Westtrate, J.A., and Tijburg, L.B.M. 1998. Bioavailability of catechins from tea: the effect of milk, *Eur. J. Clin. Nutr.*, 52:356–359.

Whitehead, T.P., Robinson, D., Allaway, S., Syms, J., and Hale, A. 1995. Effect of red wine ingestion on the antioxidant capacity of serum, *Clin. Chem.*, 41:32–35.

Wilson, T., Knight, T.J., Beitz, D.C., Lewis, D.S., and Engen, R.L. 1996. Resveratrol promotes atherosclerosis in hypercholesterolemic rabbits, *Life Sciences,* 59:15–21.

Xu, R., Yokoyama, W.H., Irving, D., Rein, D., Walzem, R.L., and German, J.B. 1998. Effect of dietary catechin and vitamin E on aortic fatty streak accumulation in hypercholesterolemic hamsters, *Atherosclerosis*, 137:29–36.

Yamakoshi, J., Kataoka, S., Koga, T., and Ariga, T. 1999. Proanthocyanidin–rich extract from grape seeds attenuates the development of aortic atherosclerosis in cholesterol-fed rabbits, *Atherosclerosis*, 142:139–149.

Zakhari, S. and Gordis, E. 1999. Moderate drinking and cardiovascular health, *Proc. Assoc. Am. Phys.*, 111:148–158.

CHAPTER 5

Beyond α-Tocopherol: The Role of the Other Tocopherols and Tocotrienols

ANDREAS M. PAPAS

CONTENTS

1-58716-083-8/02/$0.00 + $1.50
© 2002 by CRC Press LLC

INTRODUCTION

Eight natural compounds, four tocopherols, designated as α, β, γ, and δ, and four tocotrienols, also designated as α, β, γ, and δ, have vitamin E activity. Yet α-tocopherol has become synonymous with vitamin E because it is the predominant form in human and animal tissues. In addition, it is the most bioactive form based on the rat fetal resorption test, which is the classical assay for vitamin E activity. Recent research, however, shows that the other tocopherols and tocotrienols have important and unique antioxidant and other biological effects in nutrition and health and are now receiving increased attention (Papas, 1999).

This chapter will review the absorption, transport, metabolism, and biological function of tocopherols (with emphasis on non-α-tocopherols) and tocotrienols and their role in health and disease.

CHEMISTRY AND OCCURRENCE IN FOODS

Tocopherols consist of a chroman ring and a long, saturated phytyl chain. The four tocopherols differ only in the number and position of the methyl groups on the chroman ring. Tocotrienols have identical chroman rings to the corresponding toco-pherols, but their side chain is unsaturated with double bonds in the 3′, 7′, and 11′ positions (Figure 5.1). Tocopherols are found most abundantly in the oils extracted from oil seeds such as soy, corn, canola, cotton, and sunflower seeds. γ-Tocopherol is the predominant tocopherol in soy and corn oils. Tocotrienols are found primarily in the oil fractions of cereal grains such as rice, barley, rye, and wheat, and the fruit of palm (Table 5.1). Commercial quantities of tocotrienols are extracted from palm oil and rice bran oil deodorizer distillates (Papas, 1998).

Only α-tocopherol is available commercially both as natural and synthetic. Tocopherols have three asymmetric carbons at the 2′, 4′, and 8′ positions. Biosynthesis of tocopherols in nature yields only the RRR stereoisomer. For exam-ple, α-tocopherol derived from natural sources is 2R,4′R,8′R-α-tocopherol. In con-trast, α-tocopherols produced by chemical synthesis, by condensing isophytol with tri-, di-, or monomethyl hydroquinone, is an equimolar racemic mixture (all-rac) of eight stereoisomers. In commercial product labeling, natural RRR-α-tocopherol is designated as d-α-tocopherol and synthetic all-rac- is labeled as dl-α-tocopherol.

The National Research Council (NRC, 1989) recommended that, for dietary pur-poses, vitamin E activity be expressed as d-α-tocopherol equivalents (α-TE; 1.0 mg RRR-α-tocopherol = 1.0 α-TE). For mixed tocopherols the NRC proposed the fol-lowing biopotencies in a-TE: β-tocopherol 0.5, γ-tocopherol 0.1, α-tocotrienol 0.3. Commercial products are usually labeled in IU (1.0 α-TE = 1.49 IU). Vitamin E activity in commercial products is computed only from its α-tocopherol content; no IU or α-TE is included from other tocopherols or tocotrienols. The biopotency of

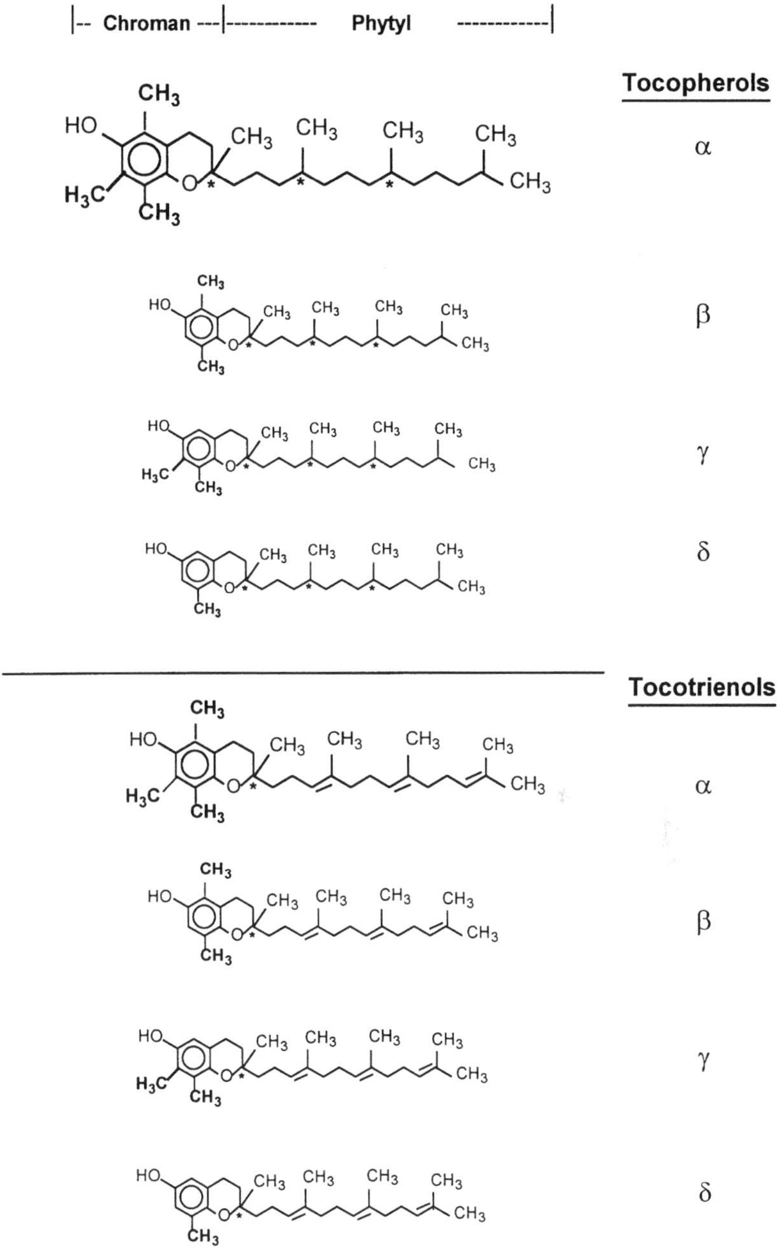

Figure 5.1 Structure of tocopherols and tocotrienols. Tocopherols and tocotrienols have the same chroman ring, but the phytyl tail of tocotrienols contains three double bonds.

TABLE 5.1
Tocopherol and tocotrienols in vegetable oils and animal fats

Oil	Tocopherols, mg/100g					Tocotrienols, mg/100g					Grand Total mg/100g
	α	β	γ	δ	Total	α	β	γ	δ	Total	
Soybean	10		59	26	96					0	96
Corn	11	5	60	2	78					0	78
Canola	17		35	1	53					0	53
Sunflower	49		5	1	55					0	55
Peanut	13		22	2	37					0	37
Cottonseed	39		39		78					0	78
Safflower	39		17	24	80					0	80
Palm	26		32	7	65	14	3	29	7	53	118
Coconut	0.5			0.6	1	0.5		2	0.6	3	4
Olive	20	1	1		22					0	22
Evening Primrose	16		42	7	65					0	65
Wheat Germ	121	65	24	25	235	2	17			19	254
Rice	12	4	5		21	18	2	57		77	98
Barley	35	5	5		45	67	12	12		91	136
Oats	18	2	5	5	30	18		3		21	51
Butter	2	0	0	0	2	0	0	0	0	0	2
Lard	1.2				1.2	0.7				0.7	1.9
Margarine	7	0	51	3	62	0	0	0	0	0	62

[1] From Papas, 1998; Papas, 1999

From Papas, A.M. (ed.), CRC Press, Boca Raton, FL. 1998; Papas, A.M. HarperCollins, New York, 1999.

synthetic dl-α-tocopherol is officially recognized as 0.74 α-TE or 1.1 IU vs. 1.0 α-TE or 1.49 IU for the naturally occurring d-α-tocopherol. Based on recent research (Burton et al., 1998; Acuff et al., 1994), the NRC recommended that the biopotency of the naturally occurring d-α-tocopherol is twice that of the synthetic dl form (Food and Nutrition Board, 2000).

ABSORPTION, TRANSPORT, AND BIOAVAILABILITY

Tocopherols are absorbed in the same path as other nonpolar lipids such as triglycerides and cholesterol (Figure 5.2). Bile, produced by the liver, emulsifies the tocopherols and incorporates them into micelles along with other fat-soluble compounds, thereby facilitating absorption. Tocopherols are absorbed from the small intestine and secreted into lymph in chylomicrons produced in the intestinal wall. Lipoprotein lipases catabolize chylomicrons rapidly and a small amount of tocopherol may be transferred from chylomicron remnants to other lipoproteins or tissues. During this process, apolipoprotein E binds to chylomicron remnants. Because the liver has specific apolipoprotein E receptors, it retains and clears the majority of the chylomicron remnants. Tocopherols in the remnants are secreted into very low-density lipoproteins (VLDL) and circulated through the plasma. VLDL is hydrolyzed by lipoprotein lipase to low density lipoproteins (LDL), which carry the largest part of plasma tocopherols and appears to exchange them readily with high-density lipoproteins (HDL). Tocopherols in HDL may be readily transferred back to chylomicron remnants as they pass through circulation returning plasma tocopherol to the liver (Kayden and Traber, 1993).

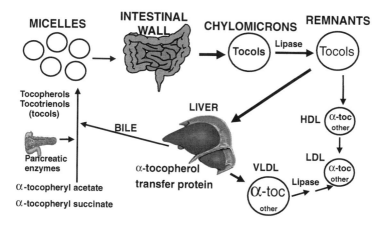

Figure 5.2 Absorption of tocopherols and tocotrienols in humans.

Absorption of tocotrienols appears similar to tocopherols. Their transport and tissue uptake, however, appear to differ from α-tocopherol. Tocotrienols disappear from plasma with chylomicron clearance and are deposited, in conjunction with triglycerides, in the adipose tissue (Hayes et al., 1993; Ikeda et al., 1996).

Our plasma and tissues contain at least 2-3 times more α- than γ-tocopherol even though the typical American diet contains more γ- than α-tocopherol (Bieri and Evarts, 1973; Murphy et al., 1990). Concentration of tocotrienols in plasma and tissues is low even when consumed in comparable amounts with α-tocopherol. All tocopherols and tocotrienols are apparently equally well absorbed. However, α-tocopherol is preferentially secreted into nascent VLDL. It was proposed that a tocopherol-binding protein is responsible for incorporating preferentially a-tocopherol into nascent VLDL. This protein has been identified and characterized in the rat, rabbit, and in humans. This protein has greater affinity for α-tocopherol than other tocopherols and tocotrienols (Kayden and Traber, 1993).

Little information is available on the catabolism of different tocopherols and tocotrienols in tissues. Two metabolic products of α-tocopherol, called the Simon metabolites, have been isolated and characterized in the urine of rabbits, humans, and rats. These are 2-(3-hydroxy-3-methyl-)-3,5,6-trimethyl-1,4-benzoquinone and its γ-lactone (Dutton et al., 1990). Two other tocopherol metabolites were isolated recently from human urine. The first, 2,5,7,8-tetramethyl-2(2′carboxyethyl)-6-hydroxychroman, a metabolite of α-tocopherol, has been suggested as an indicator of vitamin E supply (Schultz et al., 1995). The second, 2,7,8-trimethyl-2-(β-carboxyethyl)-6-hydroxychroman is believed to be a metabolite of γ-tocopherol. This metabolite, named LLU-α, has been suggested to be part of the natriuretic system, which controls extracellular fluids (Wechter et al., 1996). It was recently reported that γ-tocotrienol increases the production of LLU-α in rats (Hattori et al., 2000). It is believed that oxidized tocopherols (quinones) and other metabolic products are secreted in the bile, but information on their structure and relative abundance is very limited (Papas, 1998).

BLOOD AND TISSUE LEVELS: INTERACTIONS OF TOCOPHEROLS AND TOCOTRIENOLS

Research from several laboratories, including ours, show that high intake of α-tocopherol reduces the blood and tissue levels of γ-tocopherol (Handelman et al., 1994; Traber and Kayden, 1989; Stone et al., 1998). Since there is evidence that all tocopherols are absorbed in a similar manner (Kayden and Traber, 1993), it is possible that this effect is due to the affinity of the tocopherol transfer protein for α-tocopherol. In contrast, high levels of γ-tocopherol increase the blood levels of α-tocopherol, which may be due to a sparing effect of γ-tocopherol for α (Clement and Bourre, 1997). We observed a similar effect in our study (Figure 5.3, Stone et al., 1998). The effect of high levels of α-tocopherol on blood and tissue levels of the other tocopherols and tocotrienols is not known, although an interaction of α-tocopherol and tocotrienols has been suggested (Qureshi et al., 1996).

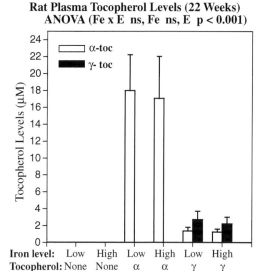

Rat Plasma Tocopherol Levels (22 Weeks)
ANOVA (Fe x E ns, Fe ns, E p < 0.001)

Figure 5.3 Plasma α- and γ-tocopherol levels after 22-week feeding. The trial included 6 dietary groups of 10 rats each. There were two levels of iron–35mg/kg (low) and 280mg/kg diet (high). Tocopherols were none, α-tocopherol (0.156 mmol/kg of feed) and γ-tocopherol (0.156 mmol/kg). Both tocopherols were in the naturally occurring RRR form. (From Stone, W. L. et al., *Cancer Detection and Prevention*, 22:S110,1998. With permission.)

The practical significance of the effect of high levels of α-tocopherol on the blood and tissue levels of γ-tocopherol (and possibly other tocopherols and tocotrienols) merits serious evaluation, especially because α-tocopherol is the predominant form used in food fortification and in dietary supplements. This is particularly important in light of the emerging understanding of the role of the other tocopherols and the tocotrienols.

TOCOPHEROLS AND TOCOTRIENOLS AS ANTIOXIDANTS

LIPID OXIDATION

A major antioxidant function of vitamin E in humans, studied mostly with α-tocopherol, is the inhibition of lipid peroxidation (Burton et al., 1982). Vitamin E also plays a critical role in preventing lipid oxidation in lipoproteins. α-Tocopherol was thought to be primarily responsible for this effect in LDL because it is the most abundant and the best scavenger of peroxyl radicals. Chylomicrons, however, carry other tocopherols and tocotrienols in similar or even higher concentration than α-tocopherol, depending on the diet, and they may also play an important role as lipid antioxidants. The other tocopherols and the tocotrienols may also function as important lipid antioxidants in adipose tissue and the liver. Some *in vitro* studies (reviewed by Papas, 1998) suggested that tocotrienols are significantly more effective than tocopherols in inhibiting LDL oxidation; in contrast, studies with plasma obtained from rats fed tocotrienol rich diets indicate approximately similar inhibition by α-tocotrienol and α-tocopherol; γ-tocopherol and γ-tocotrienol had similar effects but lower than the α forms. A recent study with rats indicated that γ-tocopherol was more effective than α-tocopherol in inhibiting LDL oxidation (Li et al., 1999; Saldeen et al., 1999). Another recent study with humans evaluated the acetate esters of the tocotrienols. The results indicated that α-tocotrienol was the most potent tocotrienol for inhibiting LDL oxidation (O'Byrne et al., 2000).

NITROGEN RADICALS

In addition to peroxy radicals, tocopherols and tocotrienols trap singlet oxygen, and other reactive species and free radicals. The antioxidant effect of vitamin E on nitrogen reactive species has been receiving increasing attention. In biological systems, nitrogen dioxide (NO_2) is produced from the reaction of nitrogen oxide (NO) with oxygen. α-Tocopherol reacts with NO_2 to yield a nitrosating agent, but this does not occur with γ-tocopherol. In contrast, γ-tocopherol reduces NO_2 to NO or reacts with it without generating a nitrosating species (Cooney et al., 1993; Christen et al., 1997). The unique ability of γ-tocopherol, probably due to the number and positioning of the methyl groups on the chroman ring, may be particularly important in carcinogenesis, arthritis, and neurologic diseases because nitrosating agents can deaminate DNA bases causing mutations or interfere with important physiological and immune functions (Wink et al., 1998; Strijbos, 1998).

OTHER ANTIOXIDANT EFFECTS

Antioxidants function as components of a complex system with significant additive, synergistic, and antagonistic effects. Additive and synergistic effects of mixtures of tocopherols have been demonstrated in food systems and *in vitro* (reviewed by Papas, 1998). Vitamin C has been shown to regenerate α-tocopherol in such systems. These interactions are very important in evaluating the overall antioxidant properties. For example, it was reported that both α- and γ-tocopherol increased superoxide dismutase (SOD) activity in plasma and arterial tissues, as well as Mn SOD and Cu/Zn SOD protein expression in arterial tissues. γ-Tocopherol was more potent than α-tocopherol in these effects (Saldeen et al., 1999).

OTHER BIOLOGICAL EFFECTS OF TOCOPHEROLS AND TOCOTRIENOLS

Tocopherols and tocotrienols and their metabolites appear to have significant and sometimes quite different metabolic effects, which may be independent or only partially related to their function as antioxidants. The following are major examples of such effects with focus on γ-tocopherol and tocotrienols. The effects of α-tocopherol were reviewed previously.

ANTI-INFLAMMATORY EFFECTS

Researchers at Berkeley evaluated the effects of α- and γ-tocopherol on γ-tocopherol reduced PGE2 synthesis in both lipopolysaccharide-stimulated macrophages and IL-1β-treated human epithelial cells. The major metabolite of dietary g-tocopherol, 2,7,8-trimethyl-2-(β-carboxyethyl)-6-hydroxychroman (γ-CEHC or LLU-α), also exhibited an inhibitory effect in these cells. In contrast, α-tocopherol slightly reduced PGE2 formation in macrophages, but had no effect in epithelial cells. The inhibitory effects of γ-tocopherol and γ-CEHC stemmed from their inhibition of COX-2 activity, rather than affecting protein expression or substrate availability, and appeared to be independent of antioxidant activity (Jiang et al., 2000).

NATRIURETIC EFFECTS?

LLU-α, a metabolite of γ-tocopherol has been suggested as an endogenous natriuretic factor (Wechter et al., 1996) which may inhibit the 70 pS K channel in the apical membrane in the kidney. It was recently reported that γ-tocotrienol increased the concentration of this metabolite in the urine of rats (Hattori et al., 2000). If the proposed function of LLU-α is confirmed, then γ-tocopherol and γ-tocotrienol may play a role in the control of extracellular fluids which is important in hypertension and cardiovascular health.

SIGNAL TRANSDUCTION

Tocopherols and tocotrienols inhibit Protein Kinase C, an important isoenzyme in signal transduction. α-Tocopherol is a stronger inhibitor than β, γ, and δ-tocopherols and α-tocotrienol (Õzer and Azzi, 1998). It was reported recently that α- and γ-tocopherol increased NO generation and nitric oxide synthase (cNOS) activity. However, only γ-tocopherol increased cNOS protein expression (Li et al., 1999). NO plays a key role in signal transduction, arterial elasticity and dilation, and has antioxidant and other major functions in the body (Ignarro et al., 1987; Murad, 1998). High levels of NO, however, can lead to production of nitrosating agent; this does not occur with g-tocopherol (Cooney et al., 1993; Christen et al., 1997). Excess NO has been implicated in cerebral ischemic neurodegeneration and excitotoxicity (Strijbos, 1998). If, indeed, γ-tocopherol stimulates the production of NO while preventing its conversion to NO_2, it could have a beneficial effect in cardiovascular health (Ignarro et al., 1987; Murad, 1998).

PLATELET ADHESION

This is the initial event in a cascade, which converts the soluble fibrinogen to insoluble fibrin and causes blood to clot. Platelet adhesion and aggregation are absolutely essential for prevention of hemorrhaging to death. Adhesion, however, of platelets to atherosclerotic lesions accelerates the formation of plaque and heart disease. α-Tocopherol and its quinone form reduce the rate of adhesion of platelets in a dose-dependent manner (Dowd and Zheng, 1995). An antioxidant effect cannot explain the activity of the quinone. Vitamin E compounds appear to modulate the function of phospholipase A_2 and lipoxygenases and production of prostacyclin, a major metabolite of arachidonic acid (Papas, 1998). It was reported recently that γ-tocopherol was more effective than α in reducing platelet aggregation and thrombus formation in rats (Saldeen et al., 1999).

EFFECTS ON CHOLESTEROL SYNTHESIS AND REMOVAL

Tocotrienols and particularly γ-tocotrienol appear to inhibit 3-hydroxy-3-methyl-glutaryl-coenzyme A (HMG-CoA) reductase *in vitro* and animal models. This enzyme is important for the synthesis of cholesterol. Specifically, it was suggested that tocotrienols inhibit the posttranscriptional suppression of HMG-CoA reductase (Parker et al., 1993). Unlike tocopherols, tocotrienols (particularly γ-tocotrienol) appear to reduce plasma apoB levels in hypercholesterolemic subjects. It has been suggested from preliminary data that γ-tocotrienol stimulates apoB degradation possibly as the result of decreased apoB translocation into the endoplasmic reticulum lumen (Theriault et al., 1999).

EFFECTS ON CANCER CELLS

Research on the role of vitamin E on cancer focused on α-tocopherol. Recent studies, primarily in cell cultures, suggest that the other tocopherols and tocotrienols may affect the growth and/or proliferation of several types of human cancer cells. Tocopherol/tocotrienol blends reduced cell growth of breast cancer cells *in vitro*, while α-tocopherol alone had no effect (Nesaretnam et al., 1998). Other researchers reported that α, γ, and δ-tocotrienols and δ-tocopherol induce breast cancer cells to undergo apoptosis (Yu et al., 1999). Another study suggested a role of γ-tocotrienol in inhibiting the growth of breast, leukemia, and melanoma cells (Mo and Elson, 1999). γ-Tocopherol appeared to be superior to α-tocopherol in inhibiting prostate cancer cells *in vitro* (Moyad et al., 1999). Our research indicated that γ-tocopherol, added to a semipurified diet, was more effective than α-tocopherol in reducing *ras*-p21 oncogenes in the colonocyte of rats (Figure 5.4, Stone et al., 1998). Although limited, the results of cancer cell studies conducted to date have been promising, and suggest a need for more research to determine the role of all tocopherols and tocotrienols (Stone and Papas, 1997).

EFFECT ON C REACTIVE PROTEIN AND CELL APOPTOSIS

It was reported recently that high levels of α-tocopherol reduced C reactive protein in healthy and diabetic people (Devaraj and Jialal, 2000). The effect of the other tocopherols and tocotrienols is not known. A recent study examined the role of γ-tocopherol in oxidized LDL-induced nuclear factor (NF)-κB activation and apopto-

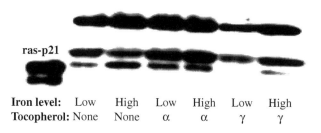

Iron level:	Low	High	Low	High	Low	High
Tocopherol:	None	None	α	α	γ	γ

Figure 5.4 Effect of α and γ-tocopherol on the expression of ras-p21 oncogenes in rat colono-cytes. The trial included 6 dietary groups of 10 rats each as described in Figure 5.3. (From Stone, W. L. et al., *Cancer Detection and Prevention*, 22:S110,1998. With permission.)

sis in human coronary artery endothelial cells. Treatment of cells with γ-tocopherol attenuated degradation of IκB and activation of NF-κB and reduced apoptosis (Li et al., 1999). Another study reported that both α- and γ-tocopherol reduced oxidized LDL-induced activation and apoptosis in human coronary artery endothelial cells *in vitro* with the α- being more effective than the γ-tocopherol (de Nigris et al., 2000). As discussed above, tocotrienols and tocopherols may induce apoptosis of some cancer cells. This would suggest a different effect of tocopherols and tocotrienols on cancerous versus noncancerous cells. It also appears that the apoptotic effects of tocopherols and tocotrienols are not uniform.

TOCOPHEROLS AND TOCOTRIENOLS IN HEALTH AND DISEASE

EFFECT ON CHOLESTEROL

Because of their effect on HMGCoA reductase, there has been significant interest on the role of tocotrienols on the lipoprotein profile. The tocotrienol rich products available commercially are mixtures of tocopherols and tocotrienols and other phytochemicals such as sterols. For this reason, direct comparisons of results from human studies with effects observed *in vitro* are difficult.

In animal studies reviewed earlier, tocotrienols and tocotrienol-rich extracts were reported to decrease total and LDL cholesterol (Watkins et al., 1998). In a recent study with apolipoprotein ApoE +/- female mice, which develop atherosclerosis only when fed diets high in triglyceride and cholesterol, mice fed a palm oil tocotrienol-rich extract had 60% lower plasma cholesterol than groups fed the atherogenic diets. Mice fed the atherogenic diet had markedly higher VLDL, intermediate density lipoprotein (IDL) and LDL cholesterol, and markedly lower HDL cholesterol than the controls. Lipoprotein patterns in mice supplemented with α-tocopherol were similar to those of the mice fed an atherogenic diet alone, but the pattern in mice supplemented with the tocotrienol extract was similar to that of mice fed the control diet (Black et al., 2000).

Human studies have produced conflicting results. In a recent study in humans, subjects were randomly assigned to receive placebo, α, γ, or δ-tocotrienyl acetate supplements (250 mg/d). Subjects followed a low-fat diet for 4 weeks, and then took supplements with dinner for the following 8 weeks while still continuing diet restrictions. α-Tocotrienyl acetate increased *in vitro* LDL oxidative resistance by 22%, and decreased its rate of oxidation. Neither serum or LDL cholesterol nor apolipoprotein B were significantly decreased by tocotrienyl acetate supplements (O'Byrne et al., 2000).

In two other studies in humans with tocotrienol-rich palm oil there was no significant effect on cholesterol profile. In the first study, conducted in the Netherlands, 20 men with slightly elevated lipid concentrations received daily for 6 weeks a palm oil-rich tocotrienol extract, which supplied 140 mg tocotrienols and 80 mg α-tocopherol; 20 other men received daily 80 mg α-tocopherol. The tocotrienol-rich extract had no marked favorable effects on the serum lipoprotein profile (Mensink et al., 2000). In the second study, a 5-year clinical study described below, a palm oil-rich tocotrienol extract which supplied approximately 240 mg tocotrienols and 98 mg α-tocopherol did not reduce cholesterol levels after 18 months' supplementation. There was, however, significant effect on LDL oxidation as indicated by thiobarbituric acid reactive substances (TBARS; Watkins et al., 1998; Tomeo et al., 1995).

In contrast, with the same experimental group, a rice bran oil tocotrienol-rich extract supplying 312 mg tocotrienols and 360 mg tocopherols significantly reduced total cholesterol, LDL, and triglycerides. HDL was higher in the treated group than in the control (Watkins et al., 1999). The conflicting results from these studies may be due in part to the levels used and the composition of the supplement, especially as it relates to other components which may have an additive or synergistic effect. In the case of rice bran oil extract there is significant literature suggesting a cholesterol-reducing effect. Sterols and other compounds present in palm oil may have a similar effect. Beneficial effects on cardiovascular health may not necessarily be associated with reduction in cholesterol levels, as discussed below.

CARDIOVASCULAR HEALTH

Epidemiological studies indicated a strong inverse association of blood vitamin E or vitamin E intake, measured as α-tocopherol, and heart disease (Rimm et al., 1993; Stampher et al., 1993). Subsequent large intervention studies with high levels of α-tocopherol failed to show the beneficial effect, especially in high-risk subjects (GISSI Group, 1999; Yusuf et al., 2000).

It is important to consider whether these results are due, at least in part, to the use of α-tocopherol alone. This is an important consideration because, as discussed above, high levels of supplemental α-tocopherol reduce the blood and tissue levels of γ-tocopherol and probably of the other tocopherols and tocotrienols. Some epidemiological studies suggest an inverse association of blood γ-tocopherol and heart disease (Kontush et al., 1999; Kristenson et al., 1997). In this regard, the role of tocotrienols may also be very important.

A 5-year study in humans evaluated 50 patients with previous cerebrovascular disease. These patients with confirmed carotid atherosclerosis were divided randomly into two groups. Twenty-five consumed a tocotrienol-rich extract which contained 240 mg of tocotrienols (plus tocopherols)/d while 25 control patients received a placebo. After 18 months, patients receiving the tocotrienol supplement demonstrated a significant regression or slowed progression in the amount of blockage of their carotid artery as revealed by ultrasonography compared to the controls (Tomeo et al., 1995). This trend continued through 4 years of supplementation. Those receiving a placebo had an overall worsening of the disease. These results are remarkable because over 40% of the patients showed regression of stenosis (Watkins, personal communication).

CANCER

A major intervention clinical study indicated that in elderly smokers receiving α-tocopherol as dl-α-tocopheryl acetate (50 mg/d), the incidence of prostate cancer was lower than in the controls (Heinonen et al., 1998). There was also a trend for lower incidence of colon cancer. Researchers at Johns Hopkins University reported that there was a stronger inverse relationship between blood γ-tocopherol and prostate cancer than the association observed with α-tocopherol (Helzlsouer et al., 2000). As discussed above, tocotrienols and δ-tocopherol inhibited the growth of several lines of cancer cells, including breast, leukemia, and melanoma. In light of these findings, it is important to evaluate the role of the tocopherols and tocotrienols in the prevention of some cancers, especially colon, breast, and prostate cancers.

OTHER CHRONIC DISEASES

The role of α-tocopherol on chronic disease has been evaluated for a variety of diseases including cataracts, Alzheimer's, and diabetes (Papas, 1999). In light of our emerging understanding of the role of the other tocopherols and tocotrienols, it is important to determine whether individually, or more important as mixtures, they have a role in the prevention and/or treatment of these diseases. This is particularly important for the γ-tocopherol, which has been suggested to affect the production and metabolism of NO which can have either beneficial or harmful effects if it is metabolized to nitrogen-reactive compounds.

CONCLUSION

Recent studies suggest that the non-α-tocopherols and the tocotrienols have antioxidant and other functions and may play an important role in nutrition and health. The understanding of their effects will help us determine whether the current practice of using α-tocopherol in food fortification and in supplements is warranted.

REFERENCES

Acuff, R. V., Thedford S. S., Hidiroglou N. N., Papas, A. M., and Odom, T. A., Jr., Relative bioavailability of RRR- and all-rac-alpha-tocopheryl acetate in humans: studies using deuterated compounds, *Am. J. Clin. Nutr.,* 1994;60:397–402.

Bieri, J. G. and Evarts R. P., Tocopherols and fatty acids in American diets, *J. Am. Diet. Assoc.* 1973;62:147–151.

Black, T. M., Wang, P., Maeda, N., and Coleman, R.A., Palm tocotrienols protect ApoE +/- mice from diet-induced atheroma formation, *J. Nutr.,* 2000;130:2420–6.

Burton, G. W., Joyce, A., and Ingold, K. U., First proof that vitamin E is major lipid-soluble, chain-breaking antioxidant in human blood plasma, *Lancet,* 1982;8292 ii:327.

Burton, G. W., Traber, M. G., Acuff, R. V., Walters, D. N., Kayden, H., Hughes, L., and Ingold, K. U., Human plasma and tissue alpha-tocopherol concentrations in response to supplementation with deuterated natural and synthetic vitamin E, *Am. J. Clin. Nutr.,* 1998;67:669–84.

Christen, S., Woodall, A. A., Shigenaga, M. K., Southwell-Keely, P. T., Duncan, M. W., and Ames, B. N., Gamma-tocopherol traps mutagenic electrophiles such as NO(X) and complements alpha-tocopherol: physiological implications, *Proc. Natl. Acad. Sci. U.S.A.,* 1997;94:3217–22.

Clement, M. and Bourre, J. M., Graded dietary levels of RRR-gamma-tocopherol induce a marked increase in the concentrations of alpha- and gamma-tocopherol in nervous tissues, heart, liver and muscle of vitamin E-deficient rats, *Biochim. Biophys. Acta.* 1997;1334:173–181.

Cooney, R. V., Franke, A. A., Harwood, P. J., Hatch-Pigott, V., Custer, L. J., and Mordan, L. J., Gamma-tocopherol detoxification of nitrogen dioxide: superiority to alpha-tocopherol, *Proc. Natl. Acad. Sci. U.S.A.,* 1993;90:1771–1775.

de Nigris, F., Franconi, F., Maida, I., Palumbo, G., Anania, V., and Napoli, C., Modulation by alpha- and gamma-tocopherol and oxidized low-density lipoprotein of apoptotic signaling in human coronary smooth muscle cells, *Biochem. Pharmacol.,* 2000;59:1477–1487.

Devaraj, S. and Jialal, I., Alpha-tocopherol supplementation decreases serum C-reactive protein and monocyte interleukin-6 levels in normal volunteers and type 2 diabetic patients, *Free Rad. Biol. Med.,* 2000;29:790–792.

Dowd, P. and Zheng Z. B., On the mechanism of the anticlotting action of vitamin E quinone, *Proc. Natl. Acad. Sci. U.S.A.,* 1995;92:8171–8175.

Dutton, P. J., Hughes, L. A., Foster, D. O., Burton, G. W., and Ingold, K.U., Simon metabolites of alpha-tocopherol are not formed via a rate-controlling scission of the 3′C-H bond, *Free Rad. Biol. Med.,* 1990;9:435–439.

Food and Nutrition Board, National Academy of Sciences, Dietary Reference Intakes for Vitamin C, Vitamin E, Selenium, and Carotenoids, National Academy Press, 2000.

GISSI Group. Dietary supplementation with n-3 polyunsaturated fatty acids and vitamin E after myocardial infarction: results of the GISSI-Prevenzione trial, *Lancet,* 1999;354:447–455.

Handelman, G. J., Epstein, W. L., Peerson, J., Spiegelman, D., and Machlin, L. J., Human adipose α-tocopherol and γ-tocopherol kinetics during and after 1 year of α-tocopherol supplementation, *Am. J. Clin. Nutr.,* 1994;59:1025–1032.

Hattori, A., Fukushima, T., Yoshimura, H., Abe, K., and Ima, K., Production of LLU-alpha following an oral administration of gamma-tocotrienol or gamma-tocopherol to rats, *Biol. Pharm. Bull.,* 2000;23:1395–1397.

Hayes, K. C., Pronczuk, A., and Liang, J. S., Differences in the plasma transport and tissue concentrations of tocopherols and tocotrienols: observations in humans and hamsters, *Proc. Soc. Exp. Biol. Med.,* 1993;202:353–359.

Heinonen, O. P., Albanes, D., and Virtamo, J., Prostate cancer and supplementation with alpha-tocopherol and beta-carotene: incidence and mortality in a controlled trial, *J. Natl. Cancer Inst.*, 1998;90:440–446.

Helzlsouer, K. J., Huang, H.-Y., Alberg, A. J., Hoffman, S., Burke, A., Norkus E. P., Morris, J. S., and Comstock, G. W., Association between alpha-tocopherol, gamma-tocopherol, selenium, and subsequent prostate cancer, *J. Natl. Cancer Inst.,* 2000:92; 2018–2023.

Ignarro, L. J., Byrns, R. E., Buga, G. M., Wood, K. S., and Chaudhuri, G., Endothelium–derived relaxing factor produced and released from artery and vein is nitric oxide, *Proc. Natl. Acad. Sci. U.S.A.*, 1987;84:9265–9269.

Ikeda, I., Imasato, Y., Sasaki, E., and Sugano, M., Lymphatic transport of alpha-, gamma-, and delta-tocotrienols and alpha-tocopherol in rats, *Int. J. Vit. Nutr. Res.,* 1996;66:217–221.

Jiang, Q., Elson-Schwab, I., Courtemanche, C., and Ames, B. N., gamma-Tocopherol and its major metabolite, in contrast to alpha-tocopherol, inhibit cyclooxygenase activity in macrophages and epithelial cells, *Proc. Natl. Acad. Sci. U.S.A.*, 2000;97:11494–11499.

Kayden, H. J. and Traber M. G., Absorption, lipoprotein transport and regulation of plasma concentrations of vitamin E in humans, *J. Lipid Res.,* 1993;34:343–358.

Kontush, A., Spranger, T., Reich, A., Baum, K., and Beisiegel, U., Lipophilic antioxidants in blood plasma as markers of atherosclerosis: the role of alpha-carotene and gamma-toco-pherol, *Atherosclerosis*, 1999;144:117–122.

Kristenson, M., Zieden, B., Kucinskiene, Z., Elinder, LS., Bergdahl, B., Elwing, B., Abaravicius, A., Razinkoviene, L., Calkauskas, H., and Olsson, A.G., Antioxidant state and mortality from coronary heart disease in Lithuanian and Swedish men: concomitant cross sectional study of men aged 50, *BMJ*, 1997;314:629–633

Li, D., Saldeen, T., Mehta, J. L., Gamma-tocopherol decreases ox-LDL-mediated activation of nuclear factor-kappaB and apoptosis in human coronary artery endothelial cells, *Biochem. Biophys. Res. Commun.*, 1999;259:157–161.

Li, D., Saldeen, T., Romeo, F., and Mehta, J. L., Relative effects of alpha- and gamma-toco-pherol on low-density lipoprotein oxidation and superoxide dismutase and nitric oxide synthase activity and protein expression in rats, *J. Cardiovasc. Pharmacol. Ther.,* 1999:219–226.

Mensink, R. P., van Houwelingen, A. C., Kromhout, D., and Hornstra, G., A vitamin E concentrate rich in tocotrienols had no effect on serum lipids, lipoproteins, or platelet function in men with mildly elevated serum lipid concentrations, *Am. J. Clin. Nutr.,* 1999;69:213–219.

Mo, H. and Elson, C. E., Apoptosis and cell-cycle arrest in human and murine tumor cells are initiated by isoprenoids, *J. Nutr.,* 1999;129:804–813.

Moyad, M. A., Brumfield, S. K., and Pienta, K. J., Vitamin E, alpha- and gamma-tocopherol, and prostate cancer, *Semin. Urol. Oncol.*, 1999;17:85–90.

Murad, F., Nitric oxide signaling: would you believe that a simple free radical could be a second messenger, autacoid, paracrine substance, neurotransmitter, and hormone? *Recent Prog. Horm. Res.*, 1998;53:43–59.

Murphy, S. P., Subar, A. F., and Block, G., Vitamin E intakes and sources in the United States, *Am. J. Clin. Nutr.,* 1990;52:361–367.

NRC. *Recommended Dietary Allowances,* 10 ed., Washington, D.C.: National Academy Press, 1989.

Nesaretnam, K., Stephen, R., Dils, R., and Darbre, P., Tocotrienols inhibit the growth of human breast cancer cells irrespective of estrogen receptor status, *Lipids*, 1998;33:461–469.

O'Byrne, D., Grundy, S., Packer, L., Devaraj, S., Baldenius, K., Hoppe, P. P., Kraemer, K., Jialal, I., and Traber M. G., Studies of LDL oxidation following alpha-, gamma-, or delta-tocotrienyl acetate supplementation of hypercholesterolemic humans, *Free Rad. Biol. Med.*, 2000;29:834–845.

Õzer, N. K. and Azzi, A., Beyond antioxidant function: other biochemical effects of antioxidants, in *Antioxidant Status, Diet, Nutrition and Health*, Papas, A.M. (ed.), CRC Press, Boca Raton, 1998;449–460.

Papas, A. M., The Vitamin E Factor, HarperCollins, New York, 1999.

Papas, A. M. (ed.), *Antioxidant Status, Diet, Nutrition and Health*, CRC Press, Boca Raton, FL, 1998.

Parker, R. A., Pearce, B. C., Clark, R. W., Gordon, D. A., and Wright, J. J., Tocotrienols regulate cholesterol production in mammalian cells by post-transcriptional suppression of 3-hydroxy-3-methylglutaryl-coenzyme A reductase, *J. Biol. Chem.*, 1993;268:11230–11238.

Qureshi, A. A., Pearce, B. C., Nor, R. M., Gapor, A., Peterson D. M., and Elson, C. E., Dietary alpha-tocopherol attenuates the impact of gamma-tocotrienol on hepatic 3-hydroxy-3-methylglutaryl coenzyme A reductase activity in chickens, *J. Nutr.*, 1996;126:389–394.

Rimm, E.B., Stampher, M. J., Ascherio, A., Giovannuci, E., Colditz, M. B., and Willet, W. C., Vitamin E consumption and the risk of coronary disease in men, *N. Engl. J. Med.*, 1993:1450–1456.

Saldeen, T., Li, D., and Mehta, J. L., Differential effects of alpha- and gamma-tocopherol on low-density lipoprotein oxidation, superoxide activity, platelet aggregation and arterial thrombogenesis, *J. Am. Coll. Cardiol.*, 1999:1208–1215.

Schultz, M., Leist, M., Petrzika, M., Gassmann, B., Brigelius-Flohé, R., A novel urinary metabolite of α-tocopherol, 2,5,7,8-tetramethyl-2(2′carboxyethyl)-6-hydroxy chroman (α-CEHC) as an indicator of an adequate vitamin E supply? *Am. J. Clin. Nutr.*, 1995;62 (suppl):1527S–1534S.

Stampfer, M. J., Hennekens, C. H., Manson, J. E., Colditz, G. A., Rosner, B., and Willett, W. C., Vitamin E consumption and the risk of coronary disease in women, *N. Engl. J. Med.*, 1993: 1444–1449.

Stone, W. L. and Papas, A. M., Tocopherols and the etiology of colon cancer, *J. Natl. Cancer. Inst.*, 1997;89:1006–1014.

Stone, W. L., Papas, A. M., LeClair, I. O., Min, Q., and Ponder, T., The influence of dietary iron and tocopherols on oxidative stress in the colon, *Cancer Detection and Prevention*, 1998, 22: S110.

Strijbos, P. J., Nitric oxide in cerebral ischemic neurodegeneration and excitotoxicity, *Crit. Rev. Neurobiol.*, 1998;12:223–243.

Theriault, A., Wang, Q., Gapor, A., and Adeli, K., Effects of gamma-tocotrienol on ApoB synthesis, degradation, and secretion in HepG2 cells, *Arterioscler. Thromb. Vasc. Biol.*, 1999;19:704–712.

Tomeo, A. C., Geller, M., Watkins, T. R., Gapor, A., and Bierenbaum, M. L., Antioxidant effects of tocotrienols in patients with hyperlipidemia and carotid stenosis, *Lipids* 1995;30:1179–1183

Traber, M. G., Kayden, H. J., Preferential incorporation of alpha-tocopherol vs. gamma-tocopherol in human lipoproteins, *Am. J. Clin. Nutr.*, 1989;49:517–526.

Watkins, T. R., Bierenbaum, M. L., and Giampaolo, A., Tocotrienols: Biological and Health Effects, in *Antioxidant Status, Diet, Nutrition and Health*, Papas, A. M. (ed.), CRC Press, Boca Raton, 1998;479–496.

Watkins, T. R., Geller, M., Kooyenga, D. K., and Bierenbaum, M., Hypocholesterolemic and antioxidant effects of rice bran oil nonsaponifiables in hypercholesterolemic subjects, *J. Environ. Nutr. Interactions,* 1999;3:115–122.

Wechter, W. J., Kantoci,. D., Murray, E. D., Jr., D'Amico, D. C., Jung, M. E., and Wang, W. H., A new endogenous natriuretic factor: LLU-alpha, *Proc. Natl. Acad. Sci. U.S.A.,* 1996;93:6002–6007.

Wink, D. A., Vodovotz, Y., Laval, J., Laval, F., Dewhirst, M. W., and Mitchell, J. B., The multifaceted roles of nitric oxide in cancer, *Carcinogenesis*, 1998;19:711–721.

Yu, W., Simmons-Menchaca, M., Gapor, A., Sanders, B. G., and Kline, K., Induction of apoptosis in human breast cancer cells by tocopherols and tocotrienols, *Nutr. Cancer,* 1999;33:26-32.

Yusuf, S., Dagenais, G., Pogue, J., Bosch, J., and Sleight, P., Vitamin E supplementation and cardiovascular events in high-risk patients, The Heart Outcomes Prevention Evaluation Study Investigators, *N. Engl. J. Med.,* 2000;342:154–160.

CHAPTER 6

Phytochemical Pharmacokinetics

HAROLD L. NEWMARK
and CHUNG S. YANG

CONTENTS

INTRODUCTION

Human beings process about 500 g of chemicals daily in the form of food. Often the bulk of this is in the form of components of plants, created by evolutionary processes for functions in the plants themselves, not necessarily designed for human intake. These plant components comprise tens of thousands of compounds in a large variety of chemical structures. Besides the (moderately) well-known proteins, carbohydrates, and fats, potent biological activities can often be found in the lesser (quantitatively) plant phenols, terpenes, terpenoids, alkaloids, purines, pyrimidines, nucleic acids, steroids, etc. Some of these compounds can produce adverse effects (e.g., β-sitosterol) or even toxic effects (e.g., digoxin), and the human body has often developed mechanisms for protection (e.g., blocking absorption of β-sitosterol). At the other end of the safety spectrum, some plant components are essential for human existence (e.g., ascorbic acid, folic acid). Often these latter few essential plant chemical components have been carefully studied for pharmacokinetic properties and con-

TABLE 6.1

Guidelines:

A. Phytochemical Pharmacokinetics Requirements
   1. Identification of pharmacologically active chemical substance(s).
   2. Determination of the nature of the pharmacologic activity, and probable mechanism(s).

B. Determination of the Active Chemical Moiety Possibilities
   1. Native plant substance as such.
   2. Food processing product (e.g., garlic allicin products)
   3. Digestive enzyme changes (e.g., esterases)
   4. Gastrointestinal chemical alterations (e.g., microbial and epithelial cell saturation of double bonds of caffeic acid and curcumin)

C. Bioavailability and Pharmacokinetics
   1. Bioavailability is generally defined as measurement of plasma concentrations after oral administration (drug, food, fluid, etc.) compared to intravenous (I.V.) administration.
   2. Few data are available on I.V. administered phytochemicals.
   3. Therefore, bioavailability of phytochemicals is often measured by plasma levels over prolonged oral intake and in comparison to a standard of reference.
   4. Pharmacokinetics can be derived from blood plasma measure ments over a short time period, usually after a single dose.

trol mechanisms in humans. However, the pharmacokinetics of most of the phytochemical components of our foods, herbs, and general environment is generally not known, or well studied.

This chapter suggests some guidelines to use in approaching studies of phytochemical pharmacokinetics, and illustrates with a few examples of active phytochemicals that have been studied, a wide variation in absorption and metabolism, resulting in potential uses and limitations in biological effects in humans (Table 6.1).

EXAMPLES

Tea, particularly the catechins of green tea, has been well studied for bioavailability, pharmacokinetics, and pharmacodynamic effect in a local tissue (e.g., lowering of prostaglandin E-2 in the human colon on oral beverage intake (Hollman et al., 1997; Yang et al., 1998; August et al., 1999)). The closely chemically related catechins of red wine also have been studied and demonstrate good bioavailability (Donovan et al., 1999). However, the circulating chemical forms of these catechins in the blood is largely as conjugates (glucuronides, sulfates), similar to other xenobiotic phenolic compounds. It is not completely clear whether biologic activity depends primarily on the circulating free phenols or whether the conjugates are also active, perhaps equally.

Quercetin, and its glycosides such as rutin, are common phenolic constituents of many plants used in human foods. Significant dietary intake is derived from onions, and tea, but lower levels are present in many foods. Both 3 and 4 hydroxyglucosides of quercetin appear equally bioavailable to humans (Olthof et al., 2000), with apparently similar pharmacokinetics.

In contrast to the catechins of tea and red wine, and quercetin and its glycosides in foods, many of the common dietary plant phenolics derived from a hydroxycinnamic acid are poorly bioavailable, due to metabolism largely within the gastrointestinal tract of humans. Caffeic acid is essentially ubiquitous in most fruits in free form but also commonly esterified with quinic acid to form chlorogenic acid (e.g., apples). Coffee beans contain about 7–8% chlorogenic acid (about 15% in dry instant coffee) (Weiss, 1957). Chlorogenic acid is rapidly hydrolyzed in the gastrointestinal tract to free caffeic acid. When 1.0 g of caffeic acid was administered orally to human volunteers, only about 11% was recovered in their urine over the following 48 h mostly as metabolites (Jacobson et al., 1983). Studies in rats previously indicated that the bulk of orally administered caffeic or chlorogenic acids is hydrogenated by intestinal microorganisms to a mixture of phenyl propionic acids which mostly appear in the feces, (not absorbed) (Booth and Williams, 1963).

Curcumin, and the closely related curcuminoids desmethoxycurcumin and bisdesmethoxy curcumin, is the yellow pigment of tumeric (spice). By itself, it is tasteless, and represents 3–5% of dried tumeric. When isolated from tumeric, curcumin has been used in ointments as a topical anti-inflammatory agent in India for centuries. In laboratory studies over the past 12 years, our group at Rutgers, and also others, have found that curcumin is a potent and effective tumor inhibitor on topical application to skin tumors of mice (Huang et al., 1992a; 1992b; 1998; Kuttan et al., 1985). It also inhibits, on oral administration, chemical carcinogen-induced tumors in the stomach (Huang et al., 1994), small intestine (Huang et al., 1994), and colon (Huang et al., 1994; Rao et al., 1995). Curcumin, extracted from tumeric and administered orally, is a very safe substance, and is therefore an FDA-approved food color. Curcumin appears to be hydrogenated to tetrahydrocurcumin, a much less effective compound, phenyl propionic acid and other metabolites, by the intestinal microorganisms and appears essentially almost completely in the feces (Huang et al., 1995).

The laboratory animal and safety studies suggest that curcumin may be a useful colon cancer chemo-preventive agent, especially since it shows inhibitory activity of cyclooxygenase and lypoxygenase enzymes affecting arachidonate metabolism (Conney et al., 1991; Huang et al., 1991). A Phase 1 clinical study in humans is currently under way (Brenner, 2000). In doses up to 0.5 g, no detectable serum levels of curcumin or tetrahydrocurcumin could be found in humans (Brenner, 2000). Thus the poor bioavailability (for absorption), and local tissue effectiveness as a colon carcinogenesis inhibitor combine to make curcumin a potentially useful, safe, and inexpensive chemopreventive agent for the colon. This, of course, must yet be demonstrated in further clinical trials.

DISCUSSION

The examples given above are just a few in the chemical class of plant phenolics. Analagous problems of the interaction of absorption, metabolism to more active agents *in vivo*, or conversely to less active metabolites and excretion, pose individual problems for phytochemicals of other chemical classes as well. An example is d-limonene which comprises over 90% of orange oil, and was found to be an inhibitory agent of tumors of the forestomach, lung, and mammary gland (Wattenberg, 1992; Gould, 1995; Elegbede et al., 1984, Haag et al., 1992). However, it was later shown that perillyl alcohol, a hydroxylated derivative (and probable metabolite) of d-limonene, was a far more active substance (Crowell et al., 1992; Haag and Gould, 1994; Mills et al., 1995). Subsequent studies in dogs and rats revealed that perillyl alcohol is rapidly absorbed from the gastrointestinal tract and metabolized to perillic acid and dihydroperillic acid, now considered to be the true active metabolites (Phillips et al., 1995).

The careful study of the pharmacokinetics of biologically active phytochemicals will likely become mandatory to prevent undue reactions, damage, and toxicity, similar to the lessons learned over the centuries from the administration of excess amounts of any xenobiotic agents. We are all familiar with the close controls required when administering digitalis alkaloids extracted from the leaves of the foxglove plant in order to achieve cardiac regulation without toxicity, controls obtained as a result of much pharmacokinetic study. A scan of some recent scientific literature strongly suggests rapidly growing requirements for many other phytochemicals as well, illustrated by the following examples:

1. Urothelial carcinoma (kidney cancer) associated with presence of *Aristolochia* species of Chinese herbs (Nortier et al., 2000).
2. Maternal oral ingestion of bioflavanoids may induce gene strand breaks and potential translocations *in utero* leading to infant and early childhood leukemia (Strick et al., 2000).
3. St. John's wort (*Hypericum perforatum*) is an herbal remedy used widely for the treatment of depression. Recent clinical studies demonstrated that *Hypericum* extracts increase the metabolism of various drugs, including combined oral contraceptives, cyclosporin, and indinavir. Moore et al., (2000) show that hyperforin, a consitutent of St. John's wort with antidepressant activity, is a potent ligand (Ki = 27 nM) for the pregnane X receptor, an orphan nuclear receptor that regulates expression of the cytochrome P450 (CYP) 3A4 monooxygenase. Treatment of primary human hepatocytes with *Hypericum* extracts or hyperforin results in a marked induction of CYP3A4 expression. Because CYP3A4 is involved in the oxidative metabolism of >

50% of all drugs, our findings provide a molecular mechanism for the interaction of St. John's wort with drugs and suggest that *Hypericum* extracts are likely to interact with many more drugs than previously had been realized.

These few examples serve to illustrate the growing importance of quantitation of intake of phytochemicals, specifically chemically identified and related to well understood biological and biochemical functions, with knowledge of their typical pharmacokinetics, as well as the range of individual deviations from the typical or normal pharmacokinetics.

CONCLUSION

In summary, phytochemicals, components of plants in our foods, herbs and general environment, comprise an enormous variety of chemical structures from many chemical classes. In practical terms, a clear definition of biological activity and structural identification of the specific phytochemical agent responsible for this activity, is the first requirement before undertaking pharmacokinetics studies. Only then can useful studies be made of characteristics and mechanisms for potentially useful manipulation to control or enhance activity. These studies would include absorption, distribution, metabolism and metabolites, effective half-life, site and mode of activity, inactivation, excretion, etc. Unfortunately, only a minority of phytochemicals with claimed biologic effects have been properly or adequately studied in this manner.

REFERENCES

August, D. A., Landau, J., Caputo, D., Hong, J., Lee, M-J., and Yang, C. S., 1999. Ingestion of Green Tea Rapidly Decreases Prostaglandin E_2 Levels in Rectal Mucosa in Humans, *Cancer Epidemiol., Biomark. Prevent.*, 8:709–713.

Booth, A. N. and Williams, R. T., 1963. Dehydroxylation of caffeic acid by rat and rabbit caecal contents and sheep rumen liquor, *Nature,* London 198:684-685.

Brenner, D., Personal communication, 2000.

Conney, A. H., Lysz, T., Ferraro, T., Abidi, T. F., Manchand, P. S., and Laskin, J. D., Huang, M-T, 1991. "Inhibitory effect of curcumin and some related dietary compounds on tumor promotion and arachidonic acid metabolism in mouse skin." in Weber, G. (ed.), *Adv. Enzyme Reg.*, vol. 31, Pergamon Press, New York, pp. 385–396.

Crowell, P .L., Kennan, W. S., Haag, J. D., Ahmad, S., Vedejs, E., and Gould, M. N., 1992. Chemoprevention of mammary carcinogenesis by hydroxylated derivatives of *d*-limonene, *Carcinogenesis*, London, 13:1261–1264.

Donovan, J. L., Bell, J. R., Kasim-Karakas, S., German J. B., Walzem, R. L., Hansen, R. J., and Waterhouse A. L., 1999. Catechin is present as metabolites in human plasma after consumption of red wine," *J. Nutr.,* 129:1662–1668.

Elegbede, J.A., Elson, C.E., Qureshi, A., Tanner, M. A., and Gould, M. N., 1984. Inhibition of DMBA-induced mammary cancer by the monoterpene *d*-limonene, *Carcinogenesis,* London, 5:661–664.

Gould, M. N., 1995. Prevention and therapy of mammary cancer by monoterpenes, *J. Cell Biochem.*, 522:139–144.

Haag, J. D. and Gould M. N., 1994. Mammary carcinoma regression induced by perillyl alcohol, a hydroxylated analog of limonene, *Cancer Chemother. Pharmacol.*, 34:477–483.

Haag, J. D., Lindstrom, M. J., and Gould, M. N., 1992. "Limonene-induced regression of mammary carcinomas," *Cancer Res.*, 52:4021–4026.

Hollman, P. C. H., Tijburg, L. B. M., and Yang, C. S., 1997. Bioavailability of Flavonoids from Tea, *Crit. Rev. Food Sci. Nutr.*, 37:719–738.

Huang, M.-T., Lou, Y.-R., Ma, W., Newmark, H. L., Reuhl, K. R., and Conney, A. H., 1994. Inhibitory effects of dietary curcumin on forestomach, duodenal, and colon carcinogenesis in mice, *Cancer Res.*, 54:5841–5847.

Huang, M.-T., Lysz T., Ferraro, T., Abidi, T. F., Laskin, J. D., and Conney, A. H., 1991. Inhibitory effects of curcumin on *in vitro* lipoxygenase and cyclooxygenase activities in mouse epidermis, *Cancer Res.*, 51:813–819.

Huang, M.-T., Ma, W., Lu, Y.-P., Chang, R. L., Fisher, C., Manchand P. S., Newmark, H. L., and Conney, A. H., 1995. Effects of curcumin, demethoxycurcumin, bisdemethoxycurcumin and tetrahydrocurcumin on 12-O-tetradecanoylphorbol-13-acetate-induced tumor promotion, *Carcinogenesis*, 16:2493–2497.

Huang, M.-T., Robertson, F. M., Lysz, T., Ferraro, T., Wang, Z. Y., Georgiadis, C. A., Laskin, J. D., and Conney, A. H., 1992a. Inhibitory effects of curcumin on carcinogenesis in mouse epidermis, in Huang, M.-T., Ho, C.-T., and Lee, C. Y., (Eds.), *Phenolic Compounds in Food and Their Effects on Health II - Antioxidants and Cancer Prevention*, Am. Chem. Soc. Symp. Ser. 507, Washington, D.C., pp. 338–349.

Huang, M.-T., Smart, R. C., Wong, C.-Q., and Conney, A. H., 1988. Inhibitory effect of curcumin, chlorogenic acid, caffeic acid, and ferulic acid on tumor promotion in mouse skin by 12-O-tetradecanoyl–phorbol-13-acetate, *Cancer Res.*, 48:5941–5946.

Huang, M.-T., Wang, Z. Y., Georgiadis, C. A., Laskin, J. D., and Conney, A. H., 1992b. Inhibitory effects of curcumin on tumor initiation by benzo[a]pyrene and 7,12-dimethylbenz[a]anthracene, *Carcinogenesis*, 13:2183-2186.

Jacobson, E., Newmark, H., Baptista, J., and Bruce, W. R., 1983. *Nutrition Reports International*, 28:1409–1417.

Kuttan, R., Bhanumathy, P., Nirmala, K., and George, M. C., 1985. Potential anticancer activity of tumeric (*Curcumal longa*), *Cancer Lett.*, 29:197–202.

Mills, J. J., Chari, R. S., Boyer, I. J., Gould, M. N., and Jirtle, R. L., 1995. Induction of apoptosis in liver tumors by the monoterpene perillyl alcohol, *Cancer Res.*, 55:979–983.

Moore, L. B., Goodwin, B., Jones, S. A., Wisely, G. B., Serabjit-Singh, C. J., Wilson, T. M., Collins, J. L., and Kliewer, S. A., 2000. St. John's wort induces hepatic drug metabolism through activation of the pregnane X receptor, *Proc. Natl. Acad. Sci. U.S.A.*, 97:7500–7502.

Nortier, J., Muniz Martinex, M.-C., Schmeiser, H. H., Arlt, V. M., Bieler, C. A., Petein, M. M., Dipierreux, M. G., De Pauw, L., Abramaowicz, D., Vereerstraeten, P., and Vanherweghem, J.-L., 2000. Urothelial carcinoma associated with the use of a Chinese herb (*Aristolochia fangchi*), *N. Engl. J. Med.*, 342:1686–1692.

Olthof, M. R., Hollman, P. C., Vree, T. B., and Katan M. B., 2000. Bioavailabilies of quercetin-3-glucoside and quercetin-4'-glucoside do not differ in humans, *J. Nutr.*, 130:1200–1203.

Phillips, L. R., Malspeis, L., and Supko, J. G., 1995. Pharmacokinetics of active drug metabolites after oral administration of perillyl alcohol, an investigational antineoplastic agent, to the dog, *Drug Metab. Dispos.*, 23:676–680.

Rao, C. V., Rivenson, A., Simi, R., and Reddy, B. S., 1995. Chemoprevention of colon carcinogenesis by dietary curcumin, a naturally occurring plant phenolic compound, *Cancer
Res.*, 55:259–266.

Strick, R., Strissel, P. L., Borgers, S., Smith, S. L., and Rowley, J. D., 2000. Dietary
bioflavonoids induce cleavage in the MLL gene and may contribute to infant leukemia,
Proc. Natl. Acad. Sci. U.S.A., 97:4790–4795.

Wattenberg, L.W. Chemoprevention of cancer by naturally occurring and synthetic compounds, in Wattenberg, L., Lipkin, M., Boone, C. W., and Kelloff, G. J., (Eds.), *Cancer
Chemoprevention*, pp. 19–49. FL, CRC Press, Boca Raton, 1992.

Weiss, L. C., 1957 (FDA Report). Report on chlorogenic acid in coffee, *J. Assoc. Off. Agric.
Chem.*, (JAOAC), 40:350–354.

Yang, C. S., Chen, L., Lee, M.-J., Balentine, D., Kuo, M. C., and Schantz, S. P., 1998. Blood
and urine levels of tea catechins after ingestion of different amounts of green tea by
human volunteers, *Cancer Epidemiol. Biomark. Prev.,* 7:351–354.

CHAPTER 7

Dietary Isothiocyanates: Roles in Cancer Prevention and Metabolism in Rodents and Humans

FUNG-LUNG CHUNG

CONTENTS

1-58716-083-8/02/$0.00 + $1.50
© 2002 by CRC Press LLC

INTRODUCTION

Isothiocyanates (ITCs) are a group of naturally occurring compounds commonly found in the crucifer family, such as watercress, broccoli, radish, and cabbage (Fenwick and Heaney, 1993). The consumption of this family of vegetables has been linked to the reduced risk of certain human cancers (Verhoeven et al., 1996; Steinmetz and Potter, 1991). It has been speculated that ITCs may contribute to the protective effects of cruciferous vegetables. The first study on the inhibition of tumor development by ITCs was published in the 1960s by Sidransky et al. (1966). In the study, it was reported that 1-naphthyl ITC, a synthetic ITC, inhibits liver tumorigenesis in rats induced by ethionine, N-2-fluorenylacetaminde, and aminoazobenzene. Almost a decade later, Wattenberg (1978) demonstrated that certain dietary aromatic ITCs suppressed benzo(a)pyrene (BaP) and 9,10-dimethyl-1,2-benzanthracene (DMBA)-induced mammary tumorigenesis in animals.

These studies prompted us to initiate a screening assay of dietary-related inhibitors against environmental nitrosamine-induced tumorigenesis in the early 1980s. The screening study marks the beginning of our long-standing effort to investigate the role of ITCs in cancer prevention. Our screening was conducted with the liver microsomes obtained from rats fed diets containing one of 25 different compounds commonly found in fruits and vegetables (Chung et al., 1984). The results demonstrated that benzyl ITC (BITC) and phenethyl ITC (PEITC), occurring primarily in gardencress and watercress, respectively, are two remarkably effective agents in blocking cytochrome P450-mediated activation of various nitrosamines, including the potent and highly specific nicotine-derived lung carcinogen, 4-(methylnitrosamino)-1-(3-pyridyl)-1-butanone (NNK) (Chung et al., 1985). These findings led us to focus on BITC and PEITC, and to test their chemopreventive potential in the lung tumor bioassays. This chapter describes our studies on the structural features and the mechanisms of ITC compounds important for tumor inhibition, as well as their metabolic fates in rodents and humans. In addition, our recent studies of sulforaphane (SFN), a major constituent of ITC in broccoli, for its effect on colon aberrant crypt foci (ACF) formation, is presented. Finally, based on analysis of a urinary marker for dietary uptake of ITCs, the potential role of ITCs in reducing lung cancer risk in humans was evaluated in a large prospective study.

ITCs AS CHEMOPREVENTIVE AGENTS FOR LUNG TUMORIGENESIS AND STRUCTURE-ACTIVITY RELATIONSHIP

Over the years we have used the A/J mouse to screen chemopreventive agents for lung tumors. The A/J mouse is a useful model, as a single dose of NNK (10 μmol) induces a significant number of lung adenomas within 16 weeks (Hecht et al., 1989). Our studies showed that the pretreatment of A/J mice with PEITC significantly inhibited the number of lung adenomas per mouse (tumor multiplicity) induced by NNK, whereas the pretreatment with lower homologues BITC and phenyl ITC

(PITC) did not affect lung tumor development (Morse et al., 1989a). A subsequent study showed that PEITC administered in the diet throughout the bioassay also inhibited lung tumorigenesis in F344 rats chronically exposed to NNK (Morse et al., 1989c).

The differences in tumor inhibitory potency by PEITC and its lower homologues in A/J mice indicate the importance of alkyl chain length. This notion was further verified by the structure-activity relationship (SAR) studies showing that the alkyl chain length is indeed a critical feature for the inhibitory activity against lung tumorigenesis, and that the potency of ITCs increase with the increased carbon chain up to six carbons (Morse et al., 1989b). Thus, the newly synthesized 6-phenylhexyl ITC (PHITC) was found to be one of the most potent inhibitors for NNK lung tumorigenesis so far studied (Chung, 1992). A closer examination of the SAR among 16 ITC compounds demonstrated that the potency of ITCs is correlated directly with lipophilicity, and inversely related to their reactivity toward glutathione (Figure 7.1) (Jiao et al., 1994). Based on SAR data, some of the synthetic diphenyl alkyl ITCs were predicted and found to have remarkable chemopreventive activity for lung tumors.

ITCs AS CHEMOPREVENTIVE AGENTS FOR TUMORIGENESIS AT OTHER ORGAN SITES

The activity of ITCs is not only limited to lung tumorigenesis, as a large number of animal studies have shown that ITCs are chemopreventive for cancer at a variety of other organ sites, including mammary gland, liver, esophagus, bladder, pancreas, and colon (Hecht, 1995; Zhang and Talalay, 1994). Most of these bioassays used protocols involving chronic administration of ITC before, during, and after the carcinogen exposure. Therefore, ITCs as a class appear to be promising and versatile agents toward prevention of chemical carcinogenesis. Among ITCs, SFN is the most abundant that occurs in broccoli and broccoli sprouts. SFN has been shown to inhibit DMBA-induced mammary tumorigenesis (Zhang et al., 1994) and, recently, AOM-induced ACF in the rat colon (Chung et al., 2000).

It should be noted, however, while results of most studies are consistent with the view that ITCs are effective inhibitors independent of the model, a few studies reported the adverse effects of the treatment with certain ITCs. For example, PEITC appears to enhance carcinogen-induced mammary and bladder tumorigenesis (Lubet et al., 1997; Hirose et al., 1998). PHITC, the synthetic compound, given in the diet, appeared to cause an increase in tumor formation in AOM-induced colon tumorigenesis (Rao et al., 1995). Even though PEITC and phenylpropyl ITC (PPITC) were remarkable inhibitors against the NMBA-induced esophageal tumor, it was found that PHITC caused an increase in tumor formation in this model (Stoner et al., 1995). These results did raise cautions in choosing the appropriate and relevant doses and models for the study of chemoprevention by ITCs, and emphasized the need for more studies to define efficacy and mechanisms.

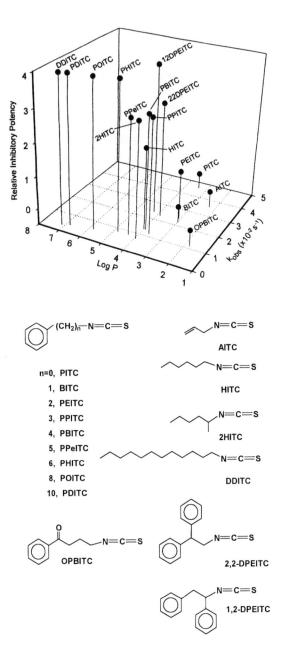

Figure 7.1 Structures of ITCs screened in lung tumorigenesis in A/J mice and the relationship of their relative potency (4 as the most potent) with lipophilicity (log P) and reactivity toward GSH (K_{obs}).

MECHANISMS OF TUMOR INHIBITION BY ITCs

Several mechanisms have been proposed and investigated for tumor inhibition by ITCs. The inhibition of NNK-induced lung tumorigenesis by PEITC and other related ITCs was mediated primarily by the inhibition of NNK metabolism. This resulted in a decrease in O^6-methylguanine in lung DNA, indicating that ITCs target cytochrome P450s (Morse et al., 1989a; Morse et al., 1991). This mechanism was verified by a number of *in vitro* and *in vivo* experiments showing specific binding and inhibition by ITCs of cytochrome p450 isozymes (Smith et al., 1996; Guo et al., 1993). ITCs such as SFN and PEITC are also potent inducers of the phase II enzymes involved in detoxification of carcinogens (Zhang and Taladay, 1994). An interesting observation was that a methylthiol conjugate of SFN, sulforamate, was found to be equally potent in phase II enzyme induction as the parent ITC, but was less toxic (Gerhauser et al., 1997). These results, together with the data of thiol conjugates of BITC and PEITC on the inhibition of lung tumorigenesis, suggest the potential of ITC conjugates as more efficacious chemopreventive agents (Jiao et al., 1997).

Phase II enzyme induction may not be critical in the prevention of nitrosamine tumorigenesis; however, it has been shown to play an important role in carcinogenesis induced by polycyclic aromatic hydrocarbons and heterocyclic aromatic amines. Many of these studies on phase I and phase II enzyme modulation constitute the early investigations on the mechanisms of ITCs. More recently, however, several laboratories demonstrated in tumor cells that ITCs, including PEITC and SFN, induce MAP kinases and other transcription factors, AP-1, c-jun, and p53, and apoptosis and arrest cell cycle (Chen et al., 1998; Huang et al., 1998; Yu et al., 1998; Gamet-Payrastre et al., 2000). These studies showed that ITC induce apoptosis and these activities may be, in part, related to the induction of phosphorylation of MAP kinases. These results thus suggest yet another potentially important mechanism by which ITCs inhibit tumorigenesis.

Results of our recent bioassay in A/J mice with PEITC and BITC and their NAC (N-acetylcysteine) conjugates administered after a single dose of either NNK or BaP appear to support the mechanisms of induction of apoptosis (Yang et al., 2002). A/J mice fed diets containing BITC or PEITC (5 µmol/g diet), or their NAC conjugates (15 µmol/g diet) for 20 weeks beginning 2 days after initiation with 20 µmol BaP showed decreases in lung tumor multiplicity by 30–40% in all treatment groups. Since ITC was given during the postinitiation stage of lung tumorigenesis, the effects cannot be attributed to the modulation of phase I or phase II enzymes.

We investigated the *in vivo* mechanisms of the postinitiation tumor inhibition by ITCs or their conjugates in BaP-treated A/J mice. Lung tissues obtained from interim sacrifices during the bioassay showed significant increases in the apoptotic index in lung tissue of BITC-NAC and PEITC-NAC groups at 84 and 140 days, with concomitant down-regulation of Bcl-2. The MAP kinase pathway was activated in lungs of treatment groups. The specific activity of JNK was detected in all treatment groups using a phosphorylation-specific antibody, with higher activity occurring in the BITC-NAC and PEITC-NAC groups. The phosphorylation level of Erk 1 was increased by PEITC-NAC and PEITC, while no significant changes in Erk 2 and p38 MAP kinase activities occurred. Downstream of MAP kinase, AP-1 and p53 were also activated. A gel shift assay showed that the AP-1 binding activity was remarkably increased in lungs from BITC-NAC and PEITC-NAC treatment groups. Phosphorylation of p53 was induced above constitutive levels after ITC treatment and was highest in PEITC-NAC and PEITC groups, but no induction of p53 accumulation was detected in any group. The study is the first *in vivo* demonstration that dietary ITCs induce MAP kinase activity, AP-1 activity, and p53 phosphorylation.

CHEMOPREVENTIVE EFFECT OF SFN AND PEITC AGAINST AOM-INDUCED ABERRANT CRYPT FOCI WITH F344 RATS

A recent case-control study in Los Angeles showed that high consumption of broccoli reduced the risk of colon cancer, and the protective effect was only found in GST-null individuals (Lin et al., 1998). Since ITCs are metabolically eliminated by GST-mediated conjugation with GSH, these authors suggested that the protection is likely to be attributed to dietary ITCs in broccoli. Surprisingly, there has been no animal data on the effect of SFN, a major ITC in broccoli, in colon tumorigenesis until recently. The scarcity of animal data on SFN has been, for the most part, due to its limited availability and cost, and yet SFN is by far the most abundant ITC (approximately 40–50% of total ITCs) in broccoli. Therefore, we carried out a bioassay to examine the effect of SFN and PEITC on AOM-induced ACF in F344 rats (Chung et al., 2000). Treatment with SFN and PEITC and their NAC conjugates by gavage (5 or 20 μmol, respectively) three doses per week for 8 weeks after AOM dosing resulted in a 30–40% reduction in formation of ACF in F344 rats (Table 7.1).

The dose of the conjugate was four times that of ITCs; however, no significant differences in the inhibition of ACF were found. These results suggest the conjugates render the inhibitory activity via gradual dissociation to parent ITCs. The protection of SFN and PEITC against ACF during the postinitiation phase of colon tumorigenesis may have important implications in the prevention of human colon cancer. However, since ACF are precancerous lesions, these results are considered preliminary. Nevertheless, the animal bioassay data are consistent with the epidemiological observations and support a potential role of SFN and PEITC in protecting against colon cancer. A preclinical efficacy study in this model is warranted.

Current data from mechanism studies also support SFN in the prevention of colon cancer. Gamet-Payrastre et al. recently showed that SFN induces cell cycle arrest in HT29 human colon cancer cells (Gamet-Payrastre et al., 2000). The growth arrest induced by SFN, followed by cell death via apoptosis, appeared to be associated with expression of cyclins A and B1. Although the mechanisms by which SFN and SFN-NAC inhibited ACF in our bioassay are yet to be investigated, its effects on apoptosis via cell cycle arrest, as demonstrated in human colon cell lines, is certainly a possibility. In addition, NAC, released by dissociation of SFN-NAC in an equilibrium, is a known antioxidant capable of inhibiting mouse fibroblast cell proliferation and locking cells in G1 phase (Sekharam et al., 1998) and induce p53-dependent

TABLE 7.1
Effects of SFN and PEITC on the formation of aberrant crypt foci induced by AOM

Treatment Group[a]	Dose of ITC Compounds (μmol)	Average Body Weight at Termination	Number of aberrant crypt foci	
			>4 Crypts	Total
1. AOM	—	310	52	153
2. AOM/ÆSFN	5	301	30 (42)[b,c]	103 (33)[d]
3. AOM/ÆSFN-NAC	20	297	31 (40)[c]	116 (24)[e]
4. AOM/ÆPEITC	5	306	27 (48)[c]	100 (35)[d]
5. AOM/ÆPEITC-NAC	20	313	38 (27)[f]	113 (26)[e]
6. SFN	5	309	0	0
7. SFN-NAC	20	312	0	0
8. PEITC	5	333	0	0
9. PEITC-NAC	20	301	0	0
10. Control	—	320	0	0

[a]ITC compounds are administered during postinitiation phase.

[b]Percent of inhibition compared to Group 1.

[c]Significantly different from Group 1 at $p < 0.01$.

[d]Significantly different from Group 1 at $p < 0.0001$.

[e]Significantly different from Group 1 at $p < 0.001$.

[f]Significantly different from Group 1 at $p < 0.05$.

apoptosis in several transformed cell lines (Liu et al., 1998). All these activities may have contributed to the inhibition of ACF in this model.

INHIBITION OF GROWTH OF HUMAN PROSTATE TUMORS BY SFN AND PEITC

Much attention has been given to prostate cancer because of its rapid increase in incidence in recent years. The notion that prostate cancer can be protected by high consumption of cruciferous vegetables has been controversial. A recent case-control study, however, showed that intake of crucifers, but not fruits or other vegetables, lowered the prostate cancer incidence (Cohen et al., 2000). We conducted a study of PEITC-NAC on LNCaP, androgen-dependent, and DU145, androgen-independent, human prostate cancer cell lines (Chiao et al., 2000). At high concentrations, PEITC-NAC caused cytolysis, while at lower concentrations PEITC-NAC mediated a dose-dependent growth modulation, with reduction of DNA synthesis and growth rate, inhibition of clonogenicity, and induction of apoptosis in both types of prostate cancer cells. PEITC-NAC decreased cells in S and G_2M phases of cell cycle, blocking cells entering replicating phases. In parallel, a significant enhancement of cells expressing the cell cycle regulator p21, as well as its intensity, was seen using a fluorescent antibody technique.

In a similar study, we also investigated the effects of SFN and its NAC conjugate (SFN-NAC) on LNCaP (Chiao et al., 2001). Both SFN and SFN-NAC mediated a dose-dependent growth inhibition and apoptosis. DNA strand breaks were detected in the apoptotic cells; total caspase activity was also elevated. SFN-NAC displayed a weaker activity than SFN in mediating apoptotic cell death. Parallel to apoptosis induction, the agents reduced the expression of cyclin D1 and the entry of G_1 cells into S and G_2M phases. DNA synthesis and subsequent cell densities were decreased in treated cell cultures. Additionally, SFN and SFN-NAC attenuated the expression of the androgen receptor and PSA production. The results indicate that SFN and SFN-NAC regulate the mechanism of cellular replication and development in human prostate cancer cells. Although tumor bioassays need to be performed, these data do suggest that dietary SFN and PEITC and their thiol conjugates may be active in the prevention of prostate cancer.

METABOLISM, TISSUE DISTRIBUTION, AND PHARMACOKINETICS OF ITCs

RODENT STUDIES

Using ^{14}C-PEITC synthesized in our laboratory with the α-carbon adjacent to the $-N=C=S$ labeled with ^{14}C, the tissue distribution and metabolism in mice treated with PEITC by gavage were studied (Eklind et al., 1990). After a single oral dose of 5 μmol (2 μCi)/mouse ^{14}C-PEITC in corn oil, a total of 50% of the dose was excreted in the urine within 48 h. Two major urinary metabolites were isolated in the urine and identified as a cyclic mercaptopyruvic acid conjugate and the NAC conjugate (Figure 7.2a). These metabolites accounted for 25 and 10%, respectively, of the administered PEITC dose.

Radioactivity in all major organs was counted, up to 72 h, after dosing of PEITC and was distributed readily in all tissues within 1 h after dosing and persisted up to 8 h. Lungs showed a maximal radioactivity between 4 and 8 h, after dosing, suggesting this time period would be optimal for inhibition. Benzyl ITC (BITC) has a similar metabolic fate as PEITC, and the main metabolite in the urine of rats dosed with BITC is the NAC conjugate (Brusewitz et al., 1977). Kassahun et al. (1997) showed that SFN is also metabolized in rats via GSH conjugation to excrete NAC conjugates in the urine as the major metabolite (>60% of the dose administered). Erucin, the sulfide analog of SFN, was produced as a metabolite, which was subsequently conjugated and excreted via mercapturic acid pathway as an NAC conjugate. These results indicate that a major metabolic pathway for ITCs through GSH conju-

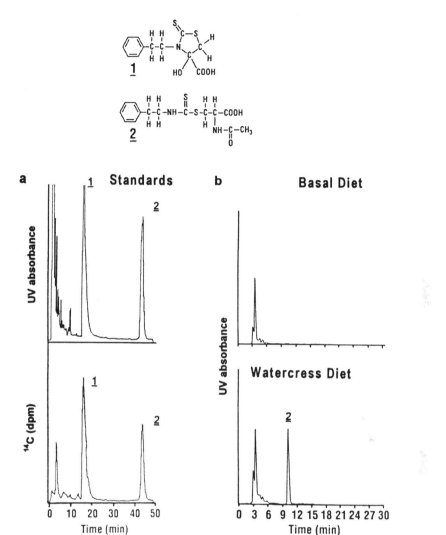

Figure 7.2 HPLC analysis of a single dose of urinary metabolites of PEITC in (a) A/J mice after oral administration of ^{14}C-PEITC. The upper panel are the UV standards of the cyclic mercaptopyruvic acid conjugate 1 and the NAC conjugate 2 and (b) humans after ingestion of watercress. The HPLC mobile phase used was different for each experiment.

gation is likely to be mediated by GST (such as GSTM1) via the mercapturic acid pathway.

We also conducted a disposition and pharmacokinetic study of PEITC and PHITC, two potent inhibitors against NNK-induced lung tumorigenesis, in F344 rats (Conaway et al., 1999). The purpose of this study was to address the question of why

PHITC is about two orders of magnitude more potent than PEITC (Morse et al., 1991). ^{14}C-PEITC and ^{14}C-PHITC were used for these studies. A single gavage dose of 50 μmol/kg (3.71 μCi/μmol) ^{14}C-PEITC or 50 μmol/kg (6.59 μCi/μmol) ^{14}C-PHITC in corn oil was administered. After ^{14}C-PEITC dosing, whole blood ^{14}C peaked at 2.9 h, with an elimination half-life of 21.7 h; blood ^{14}C from ^{14}C-PHITC-treated rats peaked at 8.9 h, with elimination half-life of 20.5 h.

In lungs, the target organ, the elimination half-life for ^{14}C-PHITC and its labeled metabolites was more than twice that for ^{14}C-PEITC and its labeled metabolites; the effective dose (area under the curve -AUC) for ^{14}C from PHITC was >2.5 times the AUC of ^{14}C from PEITC in liver, lungs, and several other tissues. During 48 h, approximately 16.5% of the administered dose of ^{14}C-PHITC was expired as ^{14}C-CO_2, more than 100 times the ^{14}C-CO_2 expired by rats treated with ^{14}C-PEITC. In rats given ^{14}C-PEITC, 88.7±2.2% and 9.9±1.9% of the dose appeared in the urine and feces, respectively, during 48 h; however, rats given ^{14}C-PHITC excreted 7.2±0.8% of the dose of ^{14}C in urine, and 47.4±14.0% in the feces. This study concluded that higher effective doses of PHITC in the lungs and other organs may be the basis, in part, for its greater potency as a chemopreventive agent.

HUMAN STUDIES

The intake of ITCs in humans is primarily through the consumption of cruciferous vegetables. When the vegetables are chewed or chopped, glucosinolates are hydrolyzed by the enzymatic action of myrosinase to yield ITC. An example of ingestion of PEITC from gluconasturtiin is shown in Figure 7.3. The metabolic fate of PEITC in humans, after ingestion of watercress, is somewhat different from that in rodents. Like mice, humans process PEITC and other ITC, such as allyl ITC from mustard, primarily by conjugation with GSH via the mercapturic acid pathway. However, humans do not produce the cyclic mercaptupyruvic acid conjugate, but only excrete the NAC conjugate in urine (Chung et al., 1992; Jiao et al., 1994) (Figure 7.2b). Approximately 50% of the PEITC administered was excreted as the NAC metabolite in the urine within 24 h. The peak of excretion was 4 h after ingestion.

Since most vegetables are consumed after being cooked, and cooking destroys myrosinase, it is important to examine whether dietary glucosinolates are actually converted to ITCs after eating cooked vegetables. A urinary marker, based on a cyclocondensation product formed by the reaction of ITCs with 1,2-benzenedithiol, was used to quantify the uptake of ITCs in humans (Zhang et al., 1992; Chung et al., 1998). Approximately one third of PEITC was excreted as PEITC-NAC after eating a total of 350 g of cooked watercress compared with uncooked watercress. These results indicate that bioavailability of PEITC is markedly compromised by cooking. The fact that the cooked watercress is completely devoid of myrosinase activity for hydrolysis of glucosinolates to ITCs and yet ITC metabolites were still found in urine, suggests that intestinal microflora are likely to be involved in converting gluco-nasturtiin to PEITC after ingesting cooked watercress (Figure 7.4a). The conversion of gluconasturtiin to PEITC upon incubation with a human fecal preparation seems to support this notion (Getahun and Chung, 1999) (Figure 7.4b).

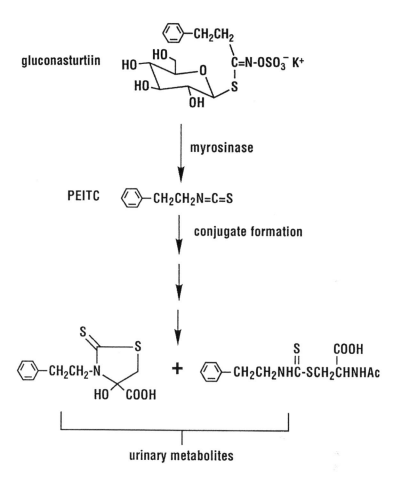

Figure 7.3 Metabolism of gluconasturtiin and PEITC in rodents and humans. Hydrolysis of gluconasturtiin by myrosinase to PEITC followed by GSH conjugation and enzymatic degradation via the mercapturic acid pathway.

Similar results on SFN were obtained from another crossover study using cooked and uncooked broccoli (Conaway et al., 2001). In the study, 12 volunteers consumed 200 g of fresh or steamed broccoli. The average 24 h urinary excretion of ITC equivalents amounted to 32.3±12.7% and 10.2±5.9% of the amounts ingested for fresh and steamed broccoli, respectively. About 40% of ITCs in urine occurred as SFN-NAC. Figure 7.5 shows the time course of total ITC excretion up to 24 h after ingestion. Not only a substantial difference in the amount of ITC excreted, but also a small shift to a later time was noted after eating steamed vs. uncooked broccoli. Shapiro et al. (1998) have drawn the same conclusions from a similar study.

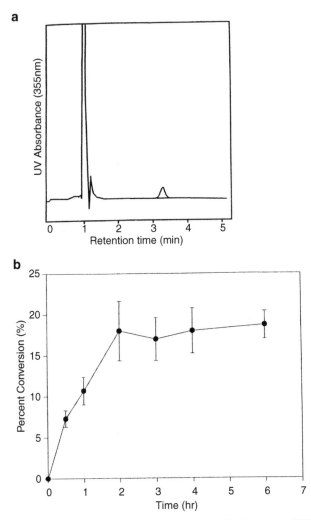

Figure 7.4   (a) HPLC analysis of a 24 h human urine sample after eating 350 g of cooked watercress; the peak at 3.2 min is ITC metabolites measured as the 1,2-benzenedithiol cyclo-condensation product; (b) conversion of glucosinolates to ITCs upon incubation of juice of cooked watercress with human fecal homogenates.

THE ROLE OF DIETARY ITCs IN THE PROTECTION AGAINST HUMAN CANCERS

There is ample evidence from animal studies supporting the potential protective effect of ITCs against cancers, yet there is little known about their roles in human cancers. It is well documented that consumption of cruciferous vegetables reduces

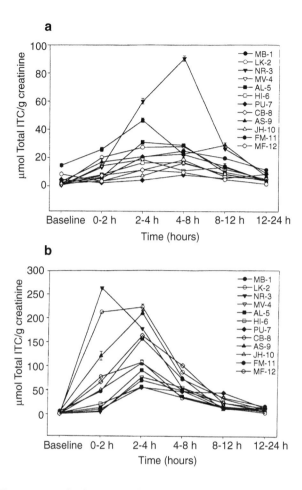

Figure 7.5 Time course of urinary excretion of ITC after ingestion of (a) steamed vs. (b) uncooked broccoli in 12 volunteers.

the risk of certain human cancers, including colorectal cancer; however, the exact nature of ingredients contributing to the beneficial effect is still not clear. There are many active compounds in cruciferous vegetables that may contribute to cancer protection.

From a chemoprevention point of view, it is important to know whether the beneficial effects of crucifers come, at least in part, from ITCs. To this end, we have collaborated with Dr. Mimi Yu and colleagues at the Norris Cancer Center of USC to apply the validated urinary marker of dietary ITCs developed in our laboratory (Seow et al., 1998). These studies have allowed us to evaluate the protective role of dietary ITCs in human cancers, such as lung, stomach, esophagus, and colon. The examination of the relationship between the amount of ITCs in urine, collected before diagnosis, of 232 lung cancer patients and 710 matched controls from a cohort of 18,244 men in Shanghai, China, followed from 1986 to 1997, has been recently completed. More than 80% of the cases are current smokers, whereas only 47% of the controls are smokers. The protective effect of dietary ITC against lung cancer was supported by the findings that individuals with detectable levels of ITCs in the urine are less likely to develop lung cancer than those with no detectable ITCs (RR,95% CI,0.60–0.65). More interesting was the observation that the reduction in risk was strongest among individuals genetically deficient in GSTM1 and T1 (RR,95% CI,0.28–0.30). Since GSTM1 is shown to be involved in conjugation of ITCs to eliminate ITCs via the mercapturic acid pathway, these results provide support for the role of ITCs in lowering risk of lung cancer (London et al., 2000).

Thus far, few studies have examined the direct relationship of ITC and human cancer. This study is the first to provide direct evidence to support the role of dietary ITCs in the prevention of human cancer, and it also suggests that the protective effect of ITCs may vary depending on the individual genetic makeup in ITC metabolism. These results warrant future clinical trials to directly evaluate the effects of ITCs on human cancers or other alternative cancer biomarkers.

CONCLUSION

ITCs are natural products that humans consume through eating cruciferous vegetables, such as watercress, broccoli, and cabbage. These compounds are readily taken up and metabolized by tissues in rodents and humans, and exhibit activities against chemical-induced carcinogenesis in various animal models. Evidence from epidemiological studies show that these compounds may play a role in the prevention of certain human cancers, a claim supported by mechanistic data from *in vitro* and *in vivo* studies. However, not everyone likes these vegetables (a notable example is former President Bush who does not like broccoli). The extensive evaluation and development of some ITCs as chemopreventive agents in clinical trials presents a practical alternative to the dietary sources.

ACKNOWLEDGMENTS

These studies would not be possible without the collaboration of many colleagues over the years. Specifically, I would like to thank Mark Morse, Karin Eklind, Stephen Hecht, Ding Jiao, Cliff Conaway, Serkadis Getahun, Shantu Amin, Bandaru Reddy, J.W. Chiao, Mimi Yu, Debra Topham, Leonard Liebes, Maria Botero-Omary, and Donald Pusateri.

REFERENCES

Brüsewitz, G., Cameron, B. D., Chasseaud, L. F., Görler, K., Hawkins, D. R., Koch, H., and Mennicke, W. H. 1977. The metabolism of benzyl isothiocyanate and its cysteine conjugate, *Biochem. J.*, 162:99–107.

Chen, Y.-R., Wang, W., Kong, A.-N. T., and Tan, T.-H. 1998. Molecular mechanisms of c-Jun N-terminal kinase-mediated apoptosis induced by anticarcinogenic isothiocyanates, *J. Biol. Chem.*, 273:1769–1775.

Chiao, J.W., Chung, F.-L., Kancherla, R., and Conaway, C.C. Sulforaphane from cruciferous vegetables and its metabolite N-acetylcysteine conjugate mediate growth arrest and apoptosis in human prostate cancer cells, *Proc. Am. Assoc. Cancer Res.*, 2001.

Chiao, J.W., Chung, F.-L, Krzeminski, J., Amin, S., Arshad, R., Ahmed, T., and Conaway, C.C. 2000. Modulation of growth of human prostate cancer cells by the N-acetylcysteine conjugate of phenethyl isothiocyanate, *Int. J. Oncol.*, 16:1215–1219.

Chung, F.-L., 1992. Chemoprevention of lung carcinogenesis by aromatic isothiocyanates, in *Cancer Chemoprevention*, Wattenberg, L., Lipkin, M., Boone, C.W., Kelloff, G.J., eds., CRC Press Inc., pp. 227–245.

Chung, F.-L., Conaway, C. C., Rao, C. V., and Reddy, B. S. 2000. Chemoprevention of colonic aberrant crypt foci in Fischer rats by sulforaphane and phenethyl isothiocyanate, *Carcinogenesis*, 21:2287-2291.

Chung, F.-L., Jiao, D., Getahun, S. M., and Yu, M. 1998. A urinary biomarker for uptake of dietary isothiocyanates in humans, *Cancer Epid. Biomark. Prev.*, 7:103–108.

Chung, F.-L., Juchatz, A., Vitarius, J., and Hecht, S. S. 1984. Effects of dietary compounds on α-hydroxylation of N-nitrosopyrrolidine and N'-nitrosonornicotine in rat target tissues, *Cancer Res.*, 44:2924–2928.

Chung, F.-L., Morse, M. A., Eklind, K. I., and Lewis, J. 1992. Quantitation of human uptake of the anticarcinogen phenethyl isothiocyanate after a watercress meal, *Cancer Epid. Biomark. Prev.*, 1, 383-388.

Chung, F.-L., Wang, M., and Hecht, S.S. 1985. Effects of dietary indoles and isothiocyanates on N-nitrosodimethylamine and 4-(methylnitrosamino)-1-(3-pyridyl)-1-butanone a-hydroxylation and DNA methylation in rat liver, *Carcinogenesis*, 6:539–543.

Cohen, J.H., Kristal, A.R., and Stanford, J. L. 2000. Fruit and vegetable intakes and prostate cancer risk, *J. Natl. Cancer Inst.*, 92:61–68.

Conaway, C. C., Getahun, S. M., Liebes, L. L., Pusateri, D. J., Topham, D. K. W., Botero-Omary, M., and Chung, F.-L., 2001. Disposition of glucosinolates and sulforaphane in humans after ingestion of steamed and fresh broccoli, *Nutrition and Cancer*, 38:168-178.

Conaway, C. C., Jiao, D., Kohri, T., Liebes, L., and Chung, F.-L. 1999. Disposition and pharmaco-kinetics of phenethyl isothiocyanate and 6-phenylhexyl isothiocyanate in F344 rats, *Drug Metab. Dispos.*, 27:13–20.

Eklind, K. I., Morse, M. A., and Chung, F.-L. 1990. Distribution and metabolism of the natural anticarcinogen phenethyl isothiocyanate in A/J mice, *Carcinogenesis*, 11:2033–2036.

Fenwick, G. R. and Heaney, R. K. 1993. Glucosinolates and their breakdown products in food and food plants, *CRC Critical Reviews in Food Science and Nutrition*, 18:123–201.

Gamet-Payrastre, L., Li, P., Lumeau, S., Cassar, G., Dupont, M.-A., Chevolleau, S., Gasc, N., Tulliez, J., and Terce, F. 2000. Sulforaphane, a naturally occurring isothiocyanate, induces cell cycle arrest and apoptosis in HT29 human colon cancer cells, *Cancer Res.*, 60:1426–1433.

Gerhauser, C., You, M., Liu, J., Moriarty, R.M., Hawthorne, M., Mehta, R. G., Moon, R. C., and Pezzuto, J. M. 1997. Cancer chemopreventive potential of sulforamate, a novel analogue of sulforaphane that induces phase 2 drug-metabolizing enzymes, *Cancer Res.*, 57:272–278.

Getahun, S. M. and Chung, F.-L. 1999. Conversion of glucosinolates in humans after ingesting cooked watercress, *Cancer Epid. Biomark. Prev.*, 8:447–451.

Guo, Z., Smith, T. J., Wang, E., Eklind, K. I., Chung, F.-L., and Yang, C. S. 1993. Structure-activity relationships of arylalkyl isothiocyanates for the inhibition of 4-(methyl-nitrosamino)-1-(3-pyridyl)-1-butanone metabolism and the modulation of xenobiotic-metabolizing enzymes in rats and mice, *Carcinogenesis*, 14:1167–1173.

Hecht, S. S. 1995. Chemoprevention by isothiocyanates, *J. Cell. Biochem. (suppl.)*, 22:195–209.

Hecht, S. S., Morse, M. A., Amin, S., Stoner, G. D., Jordan, K. G., Choi, C.-I., and Chung, F.-L. 1989. Rapid single dose model for lung tumor induction in A/J mice by 4-(methylni-trosamino)-1-(3-pyridyl)-1-butanone and the effect of diet, *Carcinogenesis*, 10:1901–1904.

Hirose, M., Yamaguchi, T., Kimoto, N., Ogawa, K., Futakuchi, M., Sano, M., and Shirai, T. 1998. Strong promoting activity of phenylethyl isothiocyanate and benzyl isothiocyanate on urinary bladder carcinogenesis in F344 male rats, *Int. J. Cancer*, 77:773–777.

Huang, C., Ma, W.-Y., Li, J., Hecht, S. S., and Dong, Z. 1998. Essential role of p53 in phenethyl isothiocyanate-induced apoptosis, *Cancer Res.*, 58:4102–4106.

Jiao, D., Eklind, K. I., Choi, C.-I., Desai, D. H., Amin, S. G., and Chung, F.-L. 1994. Structure-activity relationships of isothiocyanates as mechanism-based inhibitors of 4-(methylni-trosamino)-1-(3-pyridyl)-1-butanone-induced lung tumorigenesis in A/J mice, *Cancer Res.*, 54:4327–4333.

Jiao, D., Ho, C.-T., Foiles, P., and Chung, F.-L. 1994. Identification and quantification of the N-acetylcysteine conjugate of allyl isothiocyanate in human urine after ingestion of mustard, *Cancer Epid. Biomark. Prev.*, 3:487–492.

Jiao, D., Smith, T. J., Yang, C. S., Pittman, B., Desai, D., Amin, S., and Chung, F.-L. 1997. Chemopreventive activity of thiol conjugates of isothiocyanates for lung tumorigenesis, *Carcinogenesis*, 18:2143–2147.

Kassahun, K., Davis, M., Hu, P., Martin, B., and Baillie, T. 1997. Biotrans-formation of the naturally occurring isothiocyanate sulforaphane in the rat: identification of phase 1 metabolites and glutathione conjugates, *Chem. Res. Toxicol.*, 10:1228–1233.

Lin, H. J., Probst-Hensch, N. M., Louie, A. D., Kau, I. H., Witte, J. S., Ingles, S. A., Frankl, H. D., Lee, E. R., and Haile, R. W. 1998. Glutathione transferase null genotype, broccoli, and lower prevalence of colorectal adenomas, *Cancer Epid. Biomark. Prev.*, 7:647–652.

Liu, M., Pelling, J. C., Ju, J., Chu, E., and Brash, D. E., 1998. Antioxidant action via p53-medi-ated apoptosis, *Cancer Res.*, 58:1723–1729.

London, S. J., Yuan, J.-M, Chung, F.-L., Gao, Y.-T., Coetzee, G. A., Ross, R. K., and Yu, M. C. 2000. Isothiocyanates, glutathione S-transferase M1 and T1 polymorphisms and lung cancer risk: a prospective study of men in Shanghai, China, *Lancet*, 356:724–729.

Lubet, R. A., Steele, V. E., Eto, I., Juliana, M. M., Kelloff, G. J., and Grubbs, C. J., 1997. Chemopreventive efficacy of anethole trithione, N–acetyl-L-cysteine, miconazole and phenethylisothiocyanate in the DMBA-induced rat mammary cancer model, *Int. J. Cancer*, 72:95–101.

Morse, M. A., Amin, S. G., Hecht, S. S., and Chung, F.-L. 1989a. Effects of aromatic isothio-cyanates on tumorigenicity, O6-methylguanine formation, and metabolism of the tobacco-specific nitrosamine 4-(methylnitrosamino)-1-(3-pyridyl)-1-butanone in A/J mouse lung, *Cancer Res.*, 49:2894–2897.

Morse, M. A., Eklind, K. I., Amin, S. G., Hecht, S. S., and Chung, F.-L. 1989b. Effects of alkyl chain length on the inhibition of NNK-induced lung neoplasia in A/J mice by arylalkyl isothiocyanates, *Carcinogenesis*, 10:1757–1759.

Morse, M. A., Eklind, K. I., Hecht, S. S., Jordan, K. G., Choi, C.-I., Desai, D. H., Amin, S. G., and Chung, F.-L. 1991. Structure-activity relationships for inhibition of 4-(methylni-trosamino)-1-(3-pyridyl)-1-butanone lung tumorigenesis by arylalkyl isothiocyanates in A/J mice, *Cancer Res.*, 51:1846–1850.

Morse, M. A., Wang, C.-X., Stoner, G. D., Mandal, S., Conran, P. B., Amin, S. G., Hecht, S. S., and Chung, F.-L. 1989c. Inhibition of 4-(methylnitrosamino)-1-(3-pyridyl)-1-butanone-induced DNA adduct formation and tumorigenicity in the lung of F344 rats by dietary phenethyl isothiocyanate, *Cancer Res.*, 49:549–553.

Rao, C.V., Rivenson, A., Simi, B., Zang, E., Hamid, R., Kelloff, G. J., Steele, V., and Reddy, B. S. 1995. Enhancement of experimental colon carcinogenesis by dietary 6-phenylhexyl isothiocyanate, *Cancer Res.*, 55:4311–4318.

Sekharam, M., Trotti, A., Cunnick, J. M., and Wu, J. 1998. Suppression of fibroblast cell cycle progression in G1 phase by N-acetylcysteine, *Toxicol. Appl. Pharm.*, 149:210–216.

Seow, A., Shi, C.-Y., Chung, F.-L., Jiao, D., Hankin, J. H., Lee, H.-P., Coetzee, G. A., and Yu, M. 1998. Urinary total isothiocyanate (ITC) in a population-based sample of middle-aged and older Chinese in Singapore: Relationship with dietary total ITC and glutathione s-transferase M1/T1 genotypes, *Cancer Epid. Biomark. Prev.*, 7:775–781.

Shapiro, T. A., Fahey, J. W., Wade, K. L., Stephenson, K. K., and Talalay, P. 1998. Human metabolism and excretion of cancer chemoprotective glucosinolates and isothiocyanates of cruciferous vegetables, *Cancer Epid. Biomark. Prev.*, 7:1091–1100.

Sidransky, H., Ito, N., and Verney, E. 1966. Influence of α-naphthyl-isothiocyanate on liver tumorigenesis in rats ingesting ethionine and N-2-fluorenylacetamide, *J. Natl. Cancer Inst.*, 37:677–683.

Smith, T. J., Guo, Z., Guengerich, F. P., and Yang, C. S. 1996. Metabolism of 4-(methylni-trosamino)-1-(3-pyridyl)-1-butanone (NNK) by human cytochrome P4501A2 and its inhibition by phenethyl isothiocyanate, *Carcinogenesis*, 17:809–813.

Steinmetz, K. A. and Potter, J. D. 1991. Vegetables, fruit, and cancer. II. Mechanisms, *Cancer Causes and Control*, 2:427–442.

Stoner, G. D., Siglin, J. C., Morse, M. A., Desai, D. H., Amin, S. G., Kresty, L. A., Toburen, A. L., Heffner, E. M., and Francis, D. J. 1995. Enhancement of esophageal carcinogene-sis in male F344 rats by dietary phenylhexyl isothiocyanate, *Carcinogenesis*, 16:2473–2476.

Verhoeven, D. T. H., Goldbohm, R. A., van Poppel, G., Verhagen, H., and van den Brandt, P. A. 1996. Epidemiological studies on brassica vegetables and cancer risk, *Cancer Epid. Biomark. Prev.*, 5: 733–748.

Wattenberg, L.W. 1978. Inhibition of chemical carcinogenesis. *J. Natl. Cancer Inst.* 60:11–18.

Yang, Y.-M., Conaway, C.C., Wang, C.-X., Dai, W., Chiao, J.W., Albino, A.P., and Chung, F.-L. 2002. Mechanisms of lung tumor inhibition by isothiocyanates and their thiol conju-gates, *Cancer Res.*, in press.

Yu, R., Mandlekar, S., Harvey, K.J., Ucker, D.S., and Kong, A.-N.T. 1998. Chemopreventive isothiocyanates induce apoptosis and Caspase-3-like protease activity, *Cancer Res.*, 58:402–408.

Zhang, Y., Cho, C., Posner, G. H., and Talalay, P. 1992. Spectroscopic quantitation of organic isothiocyanates by cyclocondensation with vicinal dithiols, *Anal. Biochem.*, 205:100–107.

Zhang, Y., Kensler, T. W., Cho, C.-G., Posner, G. H., and Talalay, P. 1994. Anticarcinogenic activities of sulforaphane and structurally related synthetic norbornyl isothiocyanates, *Proc. Natl. Acad. Sci. U.S.A.,* 91:3147–3150.

Zhang, Y. and Talalay, P. 1994. Anticarcinogenic activities of organic isothiocyanates: chemistry and mechanisms, *Cancer Res. (suppl.)*, 54:1976s–1981s.

CHAPTER 8

Rationale for the Use of Soy Phytoestrogens for Neuroprotection

HELEN KIM

CONTENTS

INTRODUCTION

The purpose of this chapter is to stimulate thinking, not just about the potential benefits of the soy phytoestrogens in the brain, but also about the larger issue of the role of diet in general in determining late life health. This chapter will discuss experimental approaches taken to address whether soy phytoestrogens, or isoflavones, can have neuroprotective actions in the mammalian brain. The structural similarity between the soy isoflavones and the natural estrogen, 17β-estradiol (Figure 8.1), has been the rationale for experiments by many others that have shown that soy isoflavones can have beneficial effects in models of cardiovascular disease (Anthony et al., 1996; Clarkson et al., 1997), breast cancer, and prostate cancer (both

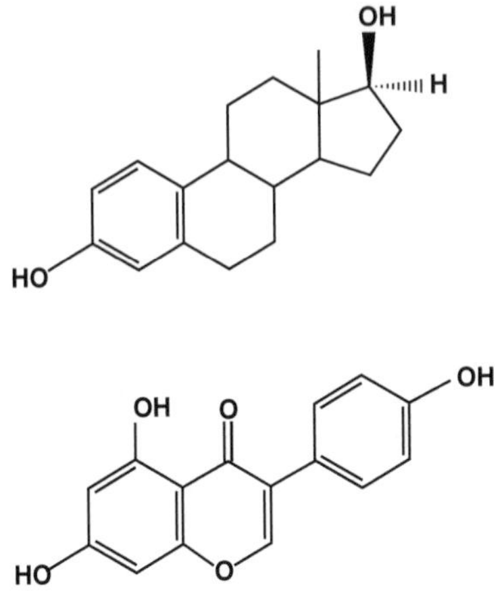

Figure 8.1 Structural similarities between the soy isoflavone genistein and 17ß-estradiol.

reviewed in Lamartiniere and Fritz, 1998). Indeed, the brain is the last frontier for testing the efficacy of soy or any phytoestrogen as an estrogen alternative, although epidemiology and experimental data provide strong rationale for both estrogen-replacement and identification of estrogen alternatives for maintaining post-menopausal brain health (Sherwin, 1988: Sherwin, 1997; McEwen and Alves, 1999; Tang et al., 1996; Toran-Allerand et al., 1999; Yaffe et al., 1998).

Our intrinsic scientific interest in the brain has been the role of the neuronal cytoskeletal elements, the microtubules, in neuronal function and viability, and the consequences of estrogen deprivation to this cytoskeletal system. Hyperphosphory-lation of the microtubule-associated protein tau has been linked with Alzheimer's disease (AD) pathology, in that the neurofibrillary tangles (NFT) that are histologic markers for AD brain are comprised of tau that is hyperphosphorylated at selected sites (Goedert et al., 1992; Kosik et al., 1988). The dogma is that although these phosphorylations are normal phosphorylations, the extent to which the hyperphos-phorylated sites are modified in tau in the NFTs renders the tau less able to associ-ate with microtubules. These in turn become more susceptible to depolymerization (Hong et al., 1998; Kim et al., 1986), leading to loss of neuronal morphology and, ultimately, function (Figure 8.2).

Thus, the identification of the molecular consequences of postmenopausal estro-gen loss is critical, particularly identification of those which are attenuated by estro-gen-replacement, or estrogen-like compounds such as the soy isoflavones. This is particularly true for estrogen loss, since it is a risk factor, not a causal factor, for AD. Obviously, not every elderly woman becomes afflicted with AD, although every woman experiences menopausal estrogen reduction. Epidemiological data obtained with postmenopausal women in the U.S. suggested, however, that even limited estro-

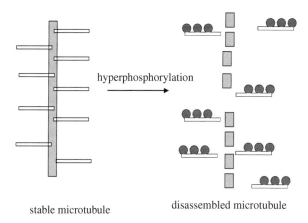

hyperphosphorylation

stable microtubule   disassembled microtubule

Figure 8.2 Schematic of the effect of hyperphosphorylation of the microtubule-associated protein tau at selected sites on microtubule stability. Neuronal function and morphology require stable microtubules, the stability of which is regulated, in part, by the associated microtubule-associated proteins, such as tau. The association of tau with microtubules is thought to be regulated by phosphorylation, which in general lowers the affinity of the tau proteins for the microtubules. Normally, these phosphorylations are thought to be reversible, and to occur at levels that maintain the neuronal microtubules in the polymeric state. In Alzheimer's disease, due to unknown molecular events, tau undergoes excessive or hyperphosphorylation at certain sites, indicated by the dots, which reduces the affinity of tau for the microtubules, resulting in the tau proteins becoming dissociated from the microtubules. The net result: a loss of microtubule integrity, and microtubule depolymerization, indicated schematically above.

gen-replacement after menopause could lower one's risk for Alzheimer's disease (Tang et al., 1996). In view of the demographics, namely that there will be a significant aged portion of the American population within our children's lifetime, any leads for reducing our collective risk for debilitating conditions such as AD are critical to pursue. Moreover, the possibilities for intervention through the use of botanical or "natural" food components such as soy, to prevent or delay the onset of late-life dementia (including AD), are crucial to explore and rigorously examine, in view of the fact that the majority of postmenopausal women resist taking estrogen-replacement therapy even given the probable adverse consequences on cardiovascular and bone health.

THE HYPOTHESIS AND EXPERIMENTS

In our laboratory, we carried out quantitative western blot analysis of homogenates of brain tissues from ovariectomized aged primates exposed to either soy isoflavones, or conjugated equine estrogens (Premarin™). The antibodies utilized recognize epitopes on the microtubule-associated protein tau, that are hyperphosphorylated in Alzheimer's disease (Binder et al., 1985; Greenberg et al., 1992). It should be emphasized that the studies are ongoing, with the full report to be published in the future.

The hypothesis underlying these experiments was that dietary intake of either Premarin™ or soy isoflavones would have neuroprotective actions which might include attenuation of AD-relevant phosphorylations of tau. Given the structural similarity between genistein, the principal soy isoflavone, and 17β-estradiol, the principal physiological estrogen (shown in Figure 8.1), it was not unreasonable to predict that the soy isoflavones might mimic the biological actions of 17β-estradiol.

The results described in this chapter were obtained from analysis of primate brain tissues archived from a study that was carried out by Dr. Tom Clarkson and his colleagues at Wake Forest University in the Department of Comparative Medicine. They set up a primate model of menopause by ovariectomizing aged *macaca fascicularis* monkeys and asked whether soy isoflavones taken through the diet were as efficacious as estrogen-replacement therapy with regard to osteoporosis and cardiovascular disease (CVD) risk factors. While part of the results of this study were published (Clarkson et al., 2001), additional data analysis is ongoing.

In brief, the ovariectomized primates were segregated into three dietary groups, all based on soy protein. One group received intact soy protein, a second group received soy protein that had been extracted of 90% of the isoflavones, and a third group received Premarin™, against the same extracted soy protein. The exact diet compositions and amounts of Premarin™ are described elsewhere (Clarkson et al., 2001). After being maintained in these dietary groups for 36 months, the monkeys were sacrificed, and their brains were dissected, sliced into 6 mm sections, frozen in liquid nitrogen, and stored at −80°C until used.

Brain samples were analyzed by quantitative western blot as follows: a piece of frontal cortex was chipped off, weighed, and pulverized under liquid nitrogen. The sample was then homogenized using a Dounce homogenizer in lysis buffer (50 mM PIPES, pH 6.9; 2 mM EGTA; 1 mM $MgCl_2$; 0.1 mM GTP, 1 tablet of Compleat™ protease inhibitor tablet [GIBCO-BRL] per 50 ml of solution) at a 1:4 g:ml ratio of tissue:buffer. This homogenate was clarified by centrifugation at 100,000 x g for 30 min. All manipulations up to this point were carried out at 4°C. The supernatant from the clarification was diluted 1:1 with 2x SDS-PAGE sample buffer (Laemmli et al., 1976) without tracking dye, and boiled at 100ºC for 5 min.

After the protein concentration was determined, 30 μg of each SDS-denatured sample were loaded on a 7.5% acrylamide gel, and electrophoresed until the tracking dye reached the bottom. The gel was then transblotted onto nitrocellulose (Towbin et al., 1979) at 130 mA overnight with cooling. Afterward, the blots were blocked for 15–30 min in 5% nonfat dry milk (NFDM) in borate buffered saline (BBS) (25 mM sodium borate, 100 mM boric acid, 75 mM NaCl), after which they were incubated in appropriately diluted primary antibody in 1% NFDM in BBS overnight with agitation. Unbound primary antibody was rinsed off with three 5 min rinses in BBS, after which the blot was agitated in peroxidase-conjugated secondary antibody diluted in 1% NFDM in BBS for 1 h at room temperature. Unbound secondary antibody was rinsed off with three 10 min rinses in BBS. Primary antibody binding was visualized with the Lumiglo chemiluminescence detection kit (Kirkegaard and Perry, Inc.) using Kodak xomat film. The film was densitometrically scanned using a BioRad GS-250 Molecular Imager, and the differences in immunoreactivity were quantitated using the Molecular Analyst software version 2.1 (BioRad).

The initial results obtained from the quantitative western blot analysis with the antibodies indicated that the brains of the animals that ingested soy protein containing the isoflavones had tau protein whose phosphorylation at two AD-relevant sites were attenuated relative to the control group which did not receive the isoflavones. Contrary to expectations, these phosphorylations were not affected in the brains of the animals that ingested Premarin™ (Kim et al., 2000). The densitometric data of the western blots are summarized in Figure 8.3.

Additional western blots with antibodies that recognized total tau proteins and total tubulin, the microtubule subunit, indicated that the amount of microtubule proteins was the same in all samples (data not shown). Thus, the differences in tau phosphorylations detected were normalized against unchanging amounts of tau and tubulin.

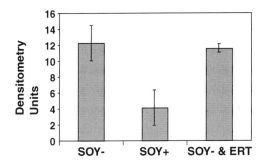

Figure 8.3 Quantitative assessment of tau phosphorylations modulated by soy versus Premarin™. Monkey brain samples were processed for SDS-PAGE and western blot as described in the text. The gel lanes were loaded with equivalent amounts of total protein (30 μg). After blocking, the blot was incubated overnight with PHF-1 antibody. (Courtesy of Dr. Sharon Greenberg, Albert Einstein College of Medicine of Yeshiva University.) Binding of primary antibody was detected with horse peroxidase-conjugated secondary antibody, and the Lumiglo chemiluminescent detection system (Kirkegaard and Perry, Inc.). After visualization, the x-ray film was scanned and the immunoreactivities for the different samples was quantitated by densitometry using Molecular Dynamics software. The graph was generated by scanning three lanes for each of the three dietary groups and averaging the intensities among the three samples in each group. Lane 1: brain homogenate from group 1 that ingested soy protein that had been extracted from the isoflavones (SOY-); lane 2: brain homogenate from group 2 that ingested soy protein that had not been extracted (SOY+); lane 3: brain homogenate from group 3 that ingested Premarin™ (conjugated equine estrogens, or CEE) added to the same soy protein that group 1 ingested. The error bars reflect the standard error of the mean.

THE RATIONALE SUPPORTING THE HYPOTHESIS AND EXPERIMENTS

This section offers a brief review of the specific evidence and understandings that led us to hypothesize that dietary soy isoflavones might have neuroprotective actions in primate brain.

ESTROGEN LOSS IS A RISK FACTOR FOR LATE-LIFE COGNITIVE DYSFUNCTION

One of the first compelling pieces of evidence that postmenopausal estrogen-replacement might protect against Alzheimer's disease in elderly women was the epidemiological study by Mayeux and coworkers (Tang et al.,1996) that showed a high correlation between estrogen use and lowered incidence of AD. Their study, which analyzed a large group of postmenopausal women in the New York City area, determined that if estrogen-replacement was undertaken for even as little as 1 year, a woman was 50% less likely to get AD. Moreover, if estrogen-replacement was undertaken for more than 1 year, a woman's risk for AD was reduced severalfold (Figure 8.4).

In related studies, animal experiments demonstrated that ovariectomy-induced estrogen-loss without replacement resulted in a measurable loss of cognitive function, that could be essentially prevented by estrogen-replacement (Singh et al., 1994; reviewed in Green and Simpkins, 2000). In considering these data however, it is crucially important to understand that while the animal studies indicate that reduction in estrogen definitely results in impairment of cognitive function, the human data demonstrated that postmenopausal estrogen-loss was a risk factor, not a causal factor, for cognitive dysfunction. Clearly other factors, as yet undefined, must play crucial roles in determining whether one woman becomes an AD patient, and another merely suffers cognitive impairment, if at all.

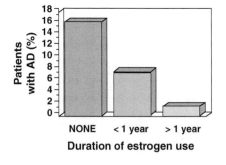

Figure 8.4 Postmenopausal estrogen replacement lowers risk for Alzheimer's disease. The three different groups of women represented here either undertook no estrogen-replacement therapy (ERT) postmenopausally, undertook ERT for up to 1 year, or undertook ERT for greater than 1 year. (Drawn from Tang, M.X. et al., 1996. *Lancet*, 348:429–432.

THE ROLE OF ESTROGEN IN PROTECTION AGAINST NEUROPATHOLOGY

It has been accepted dogma that "estrogen is good for the brain," in that the hormone was essential for the maintenance of normal neuronal viability and indeed survival (McEwen and Alves, 1999; Toran-Allerand et al.,1999). Recent experiments however, indicate a direct role for estrogen in protecting against neuropathology.

Specifically, Xu et al. (1998) demonstrated that estrogen inhibited toxic β-amyloid generation by cultured neurons. Since one of the downstream events in neurons exposed to the prooxidant β-amyloid peptide is certain pathology-relevant phosphorylations of the tau protein (Takashima et al., 1998), it was not unreasonable to postulate that the reduction in circulating estrogen levels that occurs postmenopausally may render the brain vulnerable to molecular events such as aberrantly high levels of tau phosphorylations that could lead to neuropathology or neurodegeneration.

ESTROGEN-LIKE ACTIONS OF SOY ISOFLAVONES IN MODELS OF HUMAN CHRONIC DISEASE

The correlation between estrogen loss and risk for osteoporosis and cardiovascular disease development has been strongly demonstrated by epidemiological data (Rosano and Panina, 1999). Sufficient experiments have been carried out to warrant the conclusion that dietary intake of soy protein, or specifically the soy isoflavones, results in estrogen-like protection against the development of the hormone-dependent breast and prostate cancers (Barnes, 1997; Lamartiniere and Fritz, 1998), as well as atherosclerosis (Anthony et al., 1996; Adlercreutz and Mazur, 1998; Setchell and Cassidy, 1999).

The mechanisms of protection against the cancers do not appear to be identical to those involved in cardioprotection, since the isoflavones only had effects against atherosclerosis when administered in a soy protein matrix (Clarkson et al., 1997), while breast cancer protection was measured with the soy isoflavone genistein, independent of soy protein (Murrill et al., 1996). Thus, these data foreshadowed our complex results, which suggested that the soy isoflavones and physiological estrogen have nonidentical actions in primate brain, although the end results in both cases may be beneficial, and similar behaviorally.

In terms of overall actions, it could be concluded that the soy isoflavones mimicked the effects of estrogen in protecting against the major chronic diseases that result from postmenopausal estrogen loss. These data added to the rationale for the hypothesis that dietary soy isoflavones might have neuroprotective actions in the primate model of menopause.

THE POTENTIAL ROLE OF TRANSFORMING GROWTH FACTOR BETA IN THE MECHANISM OF ACTION OF SOY ISOFLAVONES

Since its discovery, transforming growth factor beta (TGF-β) has been extensively studied for its pleiotropic actions, both in enhancing and in inhibiting cell growth (Lawrence, 1996). Because both TGF-β and genistein inhibited proliferation of cultured human cells (Roberts et al., 1990; Peterson et al., 1993), the question of whether the two factors might be linked somehow in their actions was explored, with normal human mammary epithelial (HME) cells.

First, these cells were shown to be growth inhibited at genistein concentrations attainable physiologically (Peterson et al., 1998). At concentrations of genistein that inhibited cell growth but were not cytotoxic, the resulting conditioned medium from the HME cells contained significantly higher levels of TGF-β relative to that from HME cells not exposed to genistein (Kim et al., 1998). These observations prompted the hypothesis that the cell growth-inhibitory mechanism of genistein could be mediated by TGF-β, rather than directly by the isoflavone. It is intriguing to consider that this involvement of TGF-β is an aspect of estrogen action in selected tissues: TGF-β mRNA was lowered in rats following ovariectomy (Ikeda et al.,1993), an established model for osteoporosis, where bone loss occurs predictably unless the animals are given estrogen-replacement.

A second line of evidence supports the idea that soy isoflavone action may involve TGF-β action: a limited preclinical trial of hereditary hemorrhagic telan-giactasia (HHT) patients showed that drinking a soy beverage containing soy isoflavones alleviated symptoms dramatically in several of the patients where noth-ing else had shown efficacy (Korzenik et al., 1996). HHT patients suffer chronic nasal bleeding and often require transfusions due to malformed capillary beds under-neath the skin surface (Haitjema et al., 1996; Shovlin and Letart, 1999). Of relevance to this discussion is that of the several genetic mutations that have been identified, *all* map thus far to genes that encode proteins that are involved in TGF-β signaling (McAllister et al., 1994; Vincent et al., 1995; Johnson et al., 1996; Gallione et al., 1998).

The relevance of TGF-β to the question of identifying neuroprotective actions of soy was specifically suggested by experiments that showed that TGF-β was required for serotonin-mediated long-term facilitation of synaptic connections. These experi-ments suggested that TGF-β may play an essential role in brain development, as well as in adult learning and memory (Zhang et al., 1997). Though not generally consid-ered in the context of estrogen-mimicking actions of the soy isoflavones, the poten-tially important role of TGF-β in neuronal functioning provided further rationale for examining whether dietary soy, which had been shown to have efficacy in a human disease involving mutations in TGF-β signaling, might have neuroprotective actions in a model of menopause.

DISCUSSION OF DATA OBTAINED TO DATE

As stated at the beginning of this chapter, the lab studies are ongoing, and the full analysis of the "monkey brain" data has yet to be completed. We will assume, how-ever, that the remainder of the archived monkey brains in the data set will yield data consistent with the initial numbers. What then do the data mean, and how do they extend our understanding of the actions of soy isoflavones, or of estrogens? The fact that intact soy protein, but not soy protein depleted of the isoflavones, protected against pathology-relevant tau phosphorylations implies that some component(s) in the isoflavone fraction contributed to this effect, independent of the soy protein. These data were corroborated by Pan et al. (2000), who showed that a similar soy isoflavone fraction, or 17β-estradiol, against a casein protein background, prevented ovariectomy–induced cognitive impairment in rats. An important control that Pan and his colleagues carried out was to show that neither soy isoflavones nor 17β-estradiol inhibited the beneficial actions of the other when administered together. However, again consistent with our data, the combination of soy isoflavones and 17β-estradiol did not have greater protection than either one alone, suggesting that the two were acting through complementary pathways. This latter concept may become important as experiments continue to define the actions and effects of soy and other sources of phytoestrogens on human tissues.

In brief, the data discussed here, showing different biochemical effects of soy isoflavones versus Premarin™ in the primate brain, was the first demonstration of *in vivo* modulation of tau phosphorylations. While striking, follow-up experiments are crucial to determine whether other sites of phosphorylations are affected similarly. Finally, it will be intriguing to study the differences in tau function in the brains of the monkeys in the three dietary groups. Fortunately, the function of tau can be examined *in vitro*. One of its primary functions, which is thought to mirror its *in vivo* functions, is its ability to enhance microtubule assembly when recombined with the microtubule subunit tubulin (Kim et al., 1986).

Because of the number of phosphorylations that tau can undergo, changes in function do not correlate simply with phosphorylation by a single kinase. In general, however, the phosphorylations that are hallmarks for AD are considered to reduce the affinity of tau for microtubules, which then results in microtubules being more likely to disassemble (Hong et al.,1998). It will be intriguing and important to follow up our data, to determine whether differences in function correlate with the differences in biochemical effects of soy isoflavones vs. Premarin™. Finally, in these experiments, as well as in others using similar soy protein preparations, the exact component of the soy isoflavone fraction that was responsible for the protection against AD-relevant tau phosphorylations is not understood chemically. Neither behavior nor dietary experiments have been carried out with a pure isoflavone such as genistein to determine if the beneficial effects of the intact soy protein are actually due to the isoflavones, and if so, to a single isoflavone such as genistein. Studies with pure compounds will need to be carried out to ultimately understand the molecular basis of the health benefits of soy in the primate brain.

COMPLEXITIES IN THE EXISTING LITERATURE

The data discussed in this chapter suggest that dietary intake of soy protein including the isoflavones, has health benefits in the primate brain that may protect against AD-relevant protein modifications involving the microtubule protein tau. Thus, these data suggest that incorporating soy containing isoflavones into human diets may be neuroprotective. Yet scientific data are never perfect, and in this relatively new area of research on the effects of soy in the nervous system, the data are both sparse and complex. An epidemiological study that followed the dietary patterns of a group of elderly Japanese men in Hawaii found a link between high tofu consumption and increased incidence of dementia, and reduced brain size at death (White et al., 2000). While the raw data were undeniable, the conclusion that higher tofu consumption was correlated with increased incidence of dementia is difficult to reconcile with the documented health benefits of soy and the lower incidence of dementia in Asian countries. However, until the molecular mechanisms that underlie AD pathology, and the increased risk for dementia caused by estrogen loss, are better understood, the epidemiological data of White et al. (2000) remain poorly understood, yet unchallenged.

If estrogen loss predisposed elderly women to AD, it is possible that estrogen treatment may have therapeutic effects for AD patients. But a recently completed clinical trial demonstrated that in already-diagnosed AD patients, estrogen therapy was not effective in either attenuating symptoms, or in slowing progression of the neurodegeneration, demonstrating that estrogen therapy is not a viable option for AD patients (Mulnard et al., 2000). The preponderance of the experimental data, however, predicted that estrogen treatment of already-diagnosed AD patients would not have beneficial effects. It is extremely critical to consider whether qualitatively different effects would have been obtained if a clinical trial had been carried out with women at risk for, but not yet diagnosed with, AD. Given the public health issues involved, such a clinical trial, expensive and long term though it may be, may be the only way to determine whether giving estrogen or estrogen alternatives in the premenopausal period, or pre-AD, will have efficacy in delaying the onset or incidence of AD in elderly women.

Although it appears that estrogen has no therapeutic effectiveness once neurodegeneration has been diagnosed, the concept of premenopausal treatment, particularly with a benign estrogen-alternative such as soy, to protect against a decline in cognitive function, or progression to pathology, is valid. Pan et al. (1999) showed that dietary soy isoflavones mimicked estrogen action in inhibiting ovariectomy-induced reduction in the mRNAs of critical brain proteins. More compelling, however, was the evidence that dietary soy isoflavones, against a casein protein background, protected (again like estrogen) against ovariectomy-induced decline in cognitive function in rodents (Pan et al., 2000).

Thus, while the epidemiological evidence suggested that postmenopausal estrogen-replacement therapy could lowers one's risk for AD, the clinical data strongly suggested that intervention after onset of disease was not effective (Wang et al., 2000). Limited animal behavior studies demonstrated that both estrogen and a dietary soy isoflavone fraction without soy protein protected against ovariectomy-induced cognitive dysfunction, suggesting that either estrogen or soy isoflavones, prior to the development of neurodegeneration, could be protective, as Shaywitz and Shaywitz have proposed (2000). Finally, our biochemical studies of the effects of soy vs. Premarin™ in the brain, in a primate model of menopause, indicated that soy can potentially protect against the development of pathology in the brain, but the mechanisms may be different, possibly complementary, to those of estrogen.

LOOKING INTO THE FUTURE

Projections have been made that within our children's lifetime, 1 in every 45 Americans will be living with Alzheimer's disease (Brookmeyer et al., 1998); since women live longer than men, the majority of these will be women. This will present interesting and serious public health issues, since historically women have been the primary caregivers. Taking into account the current prevalence of AD in the U.S., and everything that is known about the progression of the disease, Brookmeyer et al. (1998) projected that even a modest 1- or 2-year delay in the onset of AD will have a significant impact on the prevalence of the disease in 50 years. These projections underlie the urgent need for continued research to identify the molecular mechanisms that increase one's risk for AD and other dementias, and in turn the mechanisms that underlie potential neuroprotection by factors such as soy isoflavones and other phytochemicals.

Adding to the complexity in our understanding of human dementias, and the apparent risk factors, American and Chinese scientists collaboratively studied a large population of elderly men and women in Shanghai to determine the prevalence and incidence of dementias with respect to gender and age. They found that the patterns of dementia with respect to age groups and gender were similar to those in the U.S. An unexpected finding was that lack of education early in life among women was the most significant risk factor for late life dementia (Zhang et al., 1990). Thus, there may be a window of brain development early in life, perhaps regulated by estrogen, that may be especially dependent on environmental factors such as intellectual stimulation, so that inadequate levels of the latter have detrimental consequences for brain health much later in life. These concerns should be the basis for research.

OTHER MECHANISMS FOR NEUROPROTECTIVE ACTIONS BY SOY AND OTHER PHYTOCHEMICALS

This chapter has dealt with the potential for soy isoflavones to have potential benefit in the primate brain, in particular against brain protein modifications that are linked with AD pathology. Future experiments are important to understand both the molecular basis of human dementia and the protective actions of physiological estrogen and phytochemicals, such as the soy isoflavones, in either mimicking or complementing estrogen action. Recent experiments suggest that soy isoflavones undergo halogenation in inflammatory situations where neutrophils are stimulated to release a respiratory burst of hypochlorous acid (Boersma et al., 1999). This modification may enhance the ability of soy isoflavones to protect against oxidative damage under inflammatory conditions (Boersma et al., 1999).

Given that the brain is a primary site for aging-related oxidations, and that oxidative stress is a risk factor for AD (Joseph et al., 1998), it is important to examine protein oxidations in animal models of ovariectomy-induced cognitive dysfunction, and brains (i.e., the primate brain) from animals treated with either estrogen-replacement or soy, to determine the extent to which protein oxidations are attenuated by estrogen, and which of these are also affected by soy isoflavones. In this context, it is important to note that Joseph and coworkers showed that dietary supplementation of rodent diets with dietary plant extracts high in antioxidant activity (blueberries, spinach, strawberries) significantly protected against age-related cognitive dysfunction (Joseph et al., 1998). These studies and others (including ours) strongly suggest that certain food components may be essential for maintaining the health of the brain and other tissues, particularly late in life. Thus, modern research in "functional foods" and especially in the area of phytochemicals, may redefine "eating healthy." However, with the plethora of nonregulated botanicals-based dietary supplements currently available to the American consumer, research must define the phytochemicals that truly offer benefits, corresponding dosages and toxicity issues. Both consumers and health professionals must become better informed as to these parameters, if long-term benefits are to be truly gained from consumption of isolated phytochemicals, or even botanical extracts such as from blueberries or strawberries.

ACKNOWLEDGMENTS

During the course of the research described in this chapter, the authors were supported by funds from the United Soybean Board, a pilot grant from the NIH-funded UAB Alzheimer's Disease Research Center, the Archer Daniels Midland Company, and grant 1-P50 AT00477 (NIH Office of Dietary Supplements and the National Center for Complementary and Alternative Medicine) awarded to Purdue University, and subcontracted to the University of Alabama at Birmingham.

REFERENCES

Adlercreutz, H. and Mazur, W. (1998) Phytoestrogens and western diseases, *Korea Soybean Dig.*, 15:38–65.

Anthony, M.S., Clarkson, T.B., Bullock, B.C., and Wagner, J.D. (1996) Soy protein vs. soy phytoestrogens in the prevention of diet-induced coronary artery atherosclerosis of male cynomolgus monkeys, *Arter. Thromb. Vas. Biol.*, 17:2524–2531.

Barnes, S. (1997) The chemopreventive properties of soy isoflavonoids in animal models of breast cancer, *Breast Cancer Res. Treat.*, 46:169–179.

Binder, L.I., Frankfurter, A., and Rebhun, L.I. (1985) The distribution of tau in the mammalian central nervous system, *J. Cell Biol.*, 101:1371–1378.

Boersma, B.J., Patel, R.P., Kirk, M., Darley-Usmar, V.M., and Barnes, S. (1999) Chlorination and Nitration of Soy Isoflavones, *Arch. Biochem. Biophys.*, 368:265–275.

Brookmeyer, R., Gray, S., and Kawas, C. (1998) Projections of Alzheimer's disease in the United States and the public health impact of delaying disease onset, *Am. J. Public Health,* 88:1337–1342.

Clarkson, T.B., Anthony, M.S., Williams, J.K., Honore, E.K., and Cline, J.M. (1997) *Proc. Soc. Exper. Biol. Med.*, 217:365–368.

Clarkson, T.B., Anthony, M.S., and Morgan, T.M. (2001) Inhibition of postmenopausal atherosclerosis progression: a comparison of the effects of conjugated equine estrogens and soy phytoestrogens, *J. Clin. End. Metab.*, 86(1):41–7.

Gallione, C.J., Klaus, D.J., Yeh, E.Y., Stenzel, T.T., Xue, Y., Anthony, K.B., McAllister, K.A., Baldwin, M.A., Berg, J.N., Lux, A., Smith, J.D., Vary, C.P., Craigen, W.J., Westermann, C.J., Warner, M.L., Miller, Y.E., Jackson, C.E., Guttmacher, A.E., and Marchuk, D.A. (1998) Mutation and expression analysis of the endoglin gene in hereditary hemorrhagic telangiectasia reveals null alleles, *Hum. Mut.*, 11:286–294.

Goedert, M., Spillantini, M.G., Cairns, N.J., and Crowther, R.A. (1992) Tau proteins of Alzheimer paired helical filaments: abnormal phosphorylation of all six brain isoforms, *Neuron,* 8:159–168.

Green, P.S. and Simpkins, J.W. (2000) Estrogens and estrogen-like nonfeminizing compounds. Their role in the prevention and treatment of Alzheimer's disease, *Ann. N.Y. Acad. Sci.,* 924:93–98.

Greenberg, S.G., Davies, P., Schein, J.D., and Binder, L.I. (1992) Hydrofluoric acid-treated tau PHF proteins display the same biochemical properties as normal tau, *J. Biol. Chem.,* 267:564–569.

Haitjema, T., Westermann, C.J., Overtoom, T.T., Timer, R., Disch, F., Mauser, H., and Lammers, J.W. (1996) Hereditary hemorrhagic telangiectasia (Osler-Weber-Rendu disease): new insights in pathogenesis, complications, and treatment, *Arch. Intern. Med.,* 156:714-719.

Hong, M., Zhukareva, V., Vogelsberg-Ragaglia, V., Wszolek, Z., Reed, L., Miller, B.I., Geschwind, D.H., Bird, T.D., McKeel, D., Goate, A., Morris, J.C., Wilhelmsen, K.C., Schellenberg, G.D., Trojanowski, J.Q., and Lee, V.M. (1998) Mutation-specific functional impairments in distinct tau isoforms of hereditary FTDP-17, *Science*, 282:1914-7.

Ikeda, T., Shigeno, C., Kasai, R., Kohno, H., Ohta, S., Okumura, H., Konishi, J., and Yamamuro, T. (1993) Ovariectomy decreases the mRNA levels of transforming growth factor-beta 1 and increases the mRNA levels of osteocalcin in rat bone *in vivo*, *Biochem. Biophys. Res. Commun.*, 194:1228–1233.

Johnson, D.W., Berg, J.N., Baldwin, M.A., Gallione, C.J., Marondel, I., Yoon, S.J., Stenzel, T.T., Speer, M., Pericak-Vance, M.A., Diamond, A., Guttmacher, A.E., Jackson, C.E., Attisano, L., Kucherlapati, R., Porteous, M.E., Marchuk, D.A. (1996) Mutations in the activin receptor-like kinase 1 gene in hereditary haemorrhagic telangiectasia type 2, *Nature Gen.*, 13:189–195.

Joseph, J.A., Shukitt-Hale, B., Denisova, N.A., Prior, R.L., Cao, G., Martin, A., Taglialatela, G., and Bickford, P.C. (1998) Long-term dietary strawberry, spinach or vitamin E supplementation retards the onset of age-related neuronal signal-transduction and cognitive behavioral deficits, *J. Neurosci.,* 18:8047–8055.

Kim, H., Jensen, C.G., and Rebhun, L.I. (1986) The binding of MAP2 and tau to microtubules reassembled *in vitro*: implications for microtubule structure, *Ann. N.Y. Acad. Sci. Intl. Conf. Dynamic Aspects of Microtubule Biology,* 466:218–239.

Kim, H., Peterson, T.G., and Barnes, S. (1998) Mechanisms of action of the soy isoflavone genistein: emerging role of its effects through transforming growth factor beta signaling pathways. *Am. J. Clin. Nutr.*, 68:1418S–1425S.

Kim, H., Xia, H., Li, L., and Gewin, J. (2000) Attenuation of neurodegeneration-relevant modifications of brain proteins by dietary soy, *Biofactors*, 12:243–50.

Korzenik, J.R., Barnes, S., Coward, L., Kirk, M., and White, R.I., Jr. (1996) A pilot study of soy protein isolate in the treatment of hereditary hemorrhagic telangiectasia (HHT): possible efficacy in HHT-associated epistaxis, gastrointestinal hemorrhage and migraine. Presented at the 2nd Int. Symp. on the role of soy in the prevention and treatment of chronic disease, Brussels, Belgium.

Kosik, K.S., Orecchio, L.D, Binder, L., Trojanowski, J.Q., Lee, V.M., and Lee, G. (1988) Epitopes that span the tau molecule are shared with paired helical filaments, *Neuron*, 1:817–825.

Laemmli, U.K., Amos, L.A., and Klug, A. (1976) Correlation between structural transformation and cleavage of the major head protein of T4 bacteriophage, *Cell*, 7:191–203.

Lamartiniere, C.A. and Fritz, W.A. (1998) Genistein: chemoprevention, *in vivo* mechanisms of action, and bioavailibility, *Korea Soybean Dig.*, 15:66–80.

Lawrence, D.A. (1996) Transforming growth factor beta: a general review, *Eur. Cytokine Netw.*, 7:363–374.

McAllister, K.A., Grogg, K.M., and Johnson, D.W., et al. (1994) Endoglin, a TGFß-binding protein of endothelial cells, is the gene for hereditary haemorrhagic telangiectasia type 1, *Nat. Gen.*, 8:345–351.

McEwen B. and Alves, S. (1999) Estrogen actions in the central nervous system, *Endocr. Rev.*, 20:279–307.

Mulnard, R.A., Cotman, C., Kawas, C., van Dyck, C.H., Sano, M., Doody, R., Koss, E., Pfeiffer, E., Jin, S., Gamst, A., Grundman, M., Thomas, R., and Thal, L.J. (2000) Estrogen replacement therapy for treatment of mild to moderate Alzheimer disease: a randomized controlled trial, *JAMA*, 283:1007–1015.

Murrill, W.B., Brown, N.M., Zhang, J.-X., Manzolillo, P.A., Barnes, S., and Lamartiniere, C.A. (1996) Prepubertal genistein exposure suppresses mammary cancer and enhances gland differentiation in rats, *Carcinogenesis*, 17:1451–1457.

Pan, Y., Anthony, M., and Clarkson, T.B. (1999) Evidence for up-regulation of brain-derived neurotrophic factor mRNA by soy phytoestrogens in the frontal cortex of retired breeder female rats, *Neurosci. Lett.*, 261:1–4.

Pan, Y., Anthony, M., Watson, S., and Clarkson, T.B. (2000) Soy phytoestrogens improve radial arm maze performance in ovariectomized retired breeder rats and do not attenuate benefits of 17 beta-estradiol treatment, *Menopause*, 7(4):230–235.

Peterson, T.G. and Barnes, S. (1993) Isoflavones inhibit the growth of human prostate cancer cell lines without inhibiting epidermal growth factor receptor autophosphorylation, *Prostate*, 22:335–345.

Peterson, T.G., Ji, G.-P., Kirk, M., Coward, L., Falany, C.N., and Barnes, S. (1998) Metabolism of the isoflavones genistein and biochanin A in human breast cancer cell lines, *Am. J. Clin. Nutr.*, 68:1505–1511.

Roberts, A.B., Flanders, K.C., and Heine, U.I., et al. (1990) Transforming growth factor-beta: multifunctional regulator of differentiation and development, *Philos. Trans. R. Soc. Lond. Brit. Bio. Sci.*, 327:145–154.

Rosano, G.M. and Panina, G. (1999) Oestrogens and the heart, a review, *Therapie*, 54:381–385.

Setchell, K.D. and Cassidy, A. (1999) Dietary isoflavones: biological effects and relevance to human health, a review, *J. Nutr.*, 129:758S–767S.

Shaywitz, B.A. and Shaywitz, S.E. (2000) Estrogen and Alzheimer Disease: Plausible theory, negative clinical trial, *J. Am. Med. Assoc.*, 283:1055–1056.

Sherwin, B. (1997) Estrogen effects on cognition in menopausal women, *Neurology*, 48:S21–S26.

Sherwin, B. (1988) Estrogen and/or androgen replacement therapy and cognitive functioning in surgically menopausal women, *Psychoneuroendocrinology*, 13:345–357.

Shovlin, C.L. and Letarte, M. (1999) Hereditary haemorrhagic telangiectasia and pulmonary arteriovenous malformations: issues in clinical management and review of pathogenic mechanisms, *Thorax,* [review] 54:714–729.

Singh, M., Meyer, E.M., Millard, W.J., and Simpkins, J. (1994) Ovarian steroid deprivation results in a reversible learning impairment and compromised cholinergic function in female Soprague-Dawley rats, *Brain Res.*, 644:305–312.

Takashima, A., Honda, T., Yasutake, K., Michel, G., Murayama, O., Murayama, M., Ishiguro, K., and Yamaguchi, H. (1998) Activation of tau protein kinase I/glycogen synthase kinase-3beta by amyloid beta peptide (25-35) enhances phosphorylation of tau in hippocampal neurons, *Neurosci. Res.*, 31:317–323.

Tang, M.X., Jacobs, D., and Stern, Y., et al. (1996) Effects of oestrogen during menopause on risk and age at onset of Alzheimer's disease, *Lancet,* 348:429–432.

Toran-Allerand, C.D., Singh, M., and Setalo, G., Jr. (1999) Novel mechanisms of estrogen action in the brain: new players in an old story, a review, *Front. Neuroendocrin.*, 20:97–121.

Towbin, H., Staehelin, T., and Gordon, J. (1979) Electrophoretic transfer of proteins from polyacrylamide gels to nitrocellulose sheets: procedure and some applications, *Proc. Natl. Acad. Sci. U.S.A.*, 76:4350–4354.

Vincent, P., Plauchu, H., Hazan, J., Faure, S., Weissenbach, J., and Godet, J. (1995) A third locus for hereditary haemorrhagic telangiectasia maps to chromosome 12q. *Hum. Mol. Gen.*, 4:945–949.

Wang, P.N., Liao, S.Q., Liu, R.S., Liu, C.Y., Chao, H.T., Lu, S.R., Yu, H.Y., Wang, S.J., and Liu, H.C. (2000) Effects of estrogen on cognition, mood, and cerebral blood flow in AD: a controlled study, *Neurology*, 54:2061–2066.

White, L.R., Petrovitch, H., Ross, G.W., Masaki, K., Hardman, J., Nelson, J., Davis, D., and Markesbery, W. (2000) Brain aging and midlife tofu consumption, *J. Am. Coll. Nutr.*, 19(2):242–255.

Xu, H., Gouras, G.K., Greenfield, J.P., Vincent, B., Naslund, J., Mazzarelli, L., Fried, G., Jovanovic, J.N., Seeger, M., Relkin, N.R., Liao, F., Checler, F., Buxbaum, J.D., Chait, B.T., Thinakaran, G., Sisodia, S.S., Wang, R., Greengard, P., and Gandy, S. (1998) Estrogen reduces neuronal generation of Alzheimer beta–amyloid peptides, *Nature Med.*, 4:447–451.

Yaffe, K., Sawaya, G., Lieberburg, I., and Grady, D. (1998) Estrogen therapy in post-menopausal women: effects on cognitive function and dementia, *J. Am. Med. Assoc.*, 279:688–695.

Zhang, F., Endo, S., Cleary, L.J., Eskin, A., and Byrne, J.H. (1997) Role of transforming growth factor-beta in long-term synaptic facilitation in aplysia, *Science,* 275:1318–1320.

Zhang, M.Y., Katzman, R., Salmon, D., Jin, H., Cai, G.J., Wang. Z.Y., Qu, G.Y., Grant, I., Yu, E., and Levy, P. (1990) The prevalence of dementia and Alzheimer's disease in Shanghai, China: impact of age, gender, and education, *Ann. Neurol.*, 27:428–437.

CHAPTER 9

Bioengineering and Breeding Approaches for Improving Nutrient and Phytochemical Content of Plants

DAVID W. STILL

CONTENTS

WORLD POPULATION AND HEALTH

Feeding the world's ever-increasing population has two separate components, which are largely based on the economies of the respective country. For developing countries, simply meeting the caloric intake on a daily basis will continue to be a challenge. Because the food supply of those living in technologically advanced countries has generally been accomplished, interest has turned to developing food with greater health benefits. Plant breeders have focused most of their efforts over the previous 5 decades into increasing the yield of staple crops, and these efforts have proven wildly successful as the number of people eating less than 2100 calories per day has fallen by 75%, largely as a result of the Green Revolution (Mann, 1999).

There are a number of factors, however, that portend trouble with our food supplies. First, the majority of the world's food is provided by only 20 crop species, and it is predicted that wheat and corn will provide 80% of the caloric needs in developing countries. Because of plant domestication and the tendency for breeders to use the same or similar germplasm, the genetic diversity within each crop is surprisingly narrow (Tanksley and McCouch, 1997). Second, the maximum yields of cereal crops have not increased in 25–30 years, an indication that the genetic yield potential has been reached. Third, average yields rarely meet the maximum (record) yields. According to Kramer and Boyer (1995), average yields of all major U.S. crops are only 22% of record yields.

Since 1929, U.S. farmers have harvested approximately 96% of the acreage planted (calculated from National Agricultural Statistics Service data); 70% of the loss in yield is attributed to unfavorable physical/chemical environments (e.g., water and nutrient availability, temperature, excessive salt, etc.), while 12% is due to biotic causes (diseases, insects, and weeds). Almost 90% of the land surface in the U.S. and world-wide is subjected to physical/chemical conditions that limit crop growth (Dudal, 1976). Fourth, productive land in the U.S. and abroad is being developed for other uses that decrease biodiversity and are largely incompatible with farming. In the U.S., for example, each year for the past 40 years, 1.5% of the total acreage of farmland was lost to urbanization (U.S. Census Bureau, 2000). Past trends have shown that the land for housing and other development has largely taken place on land that was previously farmed. Fifth, the unspoken reality caused by the lack of any comprehensive local, regional, and national farmland conservation policies is that our society expects advances in agriculture and biotechnology to outpace global demand of food.

The world's population is expected to increase by over 22% to 7.5 billion people by the year 2020, with most of the population growth occurring in developing countries (U.S. Census Bureau, 2000). The WHO has predicted that by 2020 rice, wheat, and maize production will have to increase by 40% to meet the caloric needs of the world (Pinstrup-Andersen et al., 1999). Thus, there are two approaches for ensuring that the world's caloric and nutritional needs are met in the upcoming decades. The first approach is obvious, but has been adopted by governments (India and China) only after a clear crisis has presented itself, and that is to restrict population growth. The second alternative is to increase, yet again, the Earth's carrying capacity by improving the nutritional content of food, effectively delivering more nutritional value with the same, or lower, level of production. Accomplishing this will be imperative, as the genetically controlled yield potential of the major food crops has apparently been reached using current levels of breeding and biotechnology. Any further increase in yield will likely include fundamental changes in a plant's ability to add biomass by improving carbon or water use efficiency. The purpose of this article is to review the current status and future prospects of improving the nutritional content of food, including vitamins, micronutrients, and phytochemicals using plant breeding and bioengineering methods.

ROLE OF NATURAL PRODUCTS

What nutritionists and physicians define as phytochemicals are called secondary metabolites or natural products by plant biologists (Croteau et al., 2000). Classifying certain metabolites as either primary or secondary is sometimes difficult. For example, chlorophyll, while being necessary for primary metabolism, is classified as a tetraterpene, which is normally considered a secondary metabolite. In general, primary metabolites have key roles in the physiological processes central to plants, such as photosynthesis, respiration, lipid metabolism, and amino acid and nucleic acid synthesis (Table 9.1). In this paper, phytochemicals are defined as secondary metabolites, distinct from vitamins and micronutrients. Secondary metabolites are divided into three groups based on their biosynthesis: terpenoids, alkaloids, and the phenylpropanoids and associated phenolic compounds (Table 9.1). Natural products are synthesized from many of the same intermediaries, such as phosphoenolpyruvate (PEP), pyruvate, and acetyl-CoA (Figure 9.1). Plants produce more than 25,000 terpenoids, 12,000 alkaloids, and 8,000 phenolic compounds, and many secondary metabolites are unique to individual taxa. It is thought that these compounds, while not directly involved in primary plant metabolism, have evolved from interactions with other organisms: herbivores, pathogens, pollinators, and other plants. For example, plants, in response to insect herbivory, release elevated levels of volatiles, which serve as a signal to insect predators and parasitoids by the insect-damaged plants (Paré and Tumlinson, 1997). The exact function in plants of the vast majority of natural products is largely unknown and their role in human health has been provided by epidemiological studies and bioassays. In this respect, they differ from vitamins and micronutrients, whose metabolic functions and physiological consequences of deficiencies are well defined. Yet, phytochemical mining and discovery is a very active field in the medicine and pharmaceutical industry. For strategies on discovery of bioactive compounds, the reader is referred to Duke et al. (2000).

Plant biologists can manipulate the concentration of a phytochemical by one of two ways: traditional plant breeding or bioengineering. Both methods have in common the necessity to have a clear understanding of basic physiology and biochemical pathways through which plant improvement is realized. With a bioengineering approach, the biosynthetic pathway, target genes, and metabolic flux controls must be clearly elucidated. Plant breeding offers the advantage that if genetic variation exists for the chemical of interest between interbreeding species or cultivars, then the concentration of that compound can be increased in elite horticultural lines.

TABLE 9.1
Major pathways involved in primary and secondary metabolism

I. *Primary Metabolism*

A. Respiration
 - *Glycolysis* – primary pathway for the breakdown of glucose, and all carbohydrates that can be converted to glucose. Glucose is broken down to pyruvate.
 • Enzymes localize to mitochondria.
 • Phosphoenolpyruvate serves as a substrate for aromatic amino acids.
 • Pyruvate serves as a substrate for aliphatic amino acids.
 • Acetyl-CoA serves as a substrate for fatty acids and isoprene derivatives.
 - *Pentose phosphate pathway* – glucose phosphate is converted to pentose and CO_2.
 - *Tricarboxylic acid cycle* – (*Krebs cycle, citric acid cycle*) – release of reduction equivalents from activated acetate units.

B. Carbohydrate Synthesis and Metabolism
 • Carbohydrates are classified as either structural or storage polysaccharides.
 • Synthesis from carbon dioxide (*carbon-linked reactions*) or by *gluconeogenesis*.
 - *Carbon-linked reactions (Calvin cycle)* produce 3-phosphoglycerate (3PGA) or phosphophenolpyruvate (PEP), which is converted into oxaloacetate (OAA).
 • Enzymes localize to the chloroplast.
 - *Light reactions* – (water oxidations) produce O_2, ATP, and NADPH.
 • Enzymes localize to the chloroplast.
 - *Gluconeogenesis* is the net process by which lipids (triglycerides) are converted to sucrose.
 • Process occurs in the glyoxysome, mitochondria, spherosome, and the cytoplasm.

C. Lipid Metabolism
 • Lipids are classified as fatty acids, neutral lipids, and polar lipids.
 - *Glyoxylate cycle* – converts fats (acetate units) to sugars; occurs in glyoxysome, cytosol, and mitochondria.
 • Enzymes localize to plastids.
 • Acetyl-CoA is the initial substrate for synthesis of the carbon backbone of all fatty acids.

D. Amino Acid Synthesis
 - *GS/GOGAT pathway* functions as the primary assimilation of inorganic N and secondary assimilation of free ammonium.
 • Enzymes localize to plastids.
 • Synthesis of glutamine, glutamate, asparagines, aspartate.
 - *Aromatic amino acid pathways:* important precursors of primary and secondary metabolism.
 • Enzymes localize to plastids.
 • Phenylalanine and tyrosine serve as precursors of alkaloids, flavonoids, isoflavonoids, hydroxycinnamic acid, and lignin.
 • Tryptophan serves as a precursor for indole phytoalexins, indole alkaloids, and indole glucosinolates.
 - *Aspartate-derived amino acid pathway:* leads to lysine, threonine, and methionine.
 • Required in human diets.
 • Enzymes localize to chloroplasts, mainly, and the cytosol.
 • Methionine incorporated into proteins; S-adenosylmethionine used for transmethylation of lipids, pectins, chlorophyll, and nucleic acids.
 - *Branched-chain amino acids.*
 • Includes threonine, isoleucine, valine and leucine.
 • Isoleucine and valine are synthesized in chloroplasts.
 • Acetohydroxy acid synthase (AHAS) is a key enzyme in valine and isoleucine synthesis, and a target for herbicides.

E. Purine and Pyrimidine Synthesis
- *Pyrimidine nucleotides* are synthesized from the orotic acid pathway.
  • Amino donor is glutamine.
  • All enzymes used in pyrimidine synthesis localize to plastids.
- *Purine nucleotides* are synthesized directly from 5-phosphoribosyl-1-pyrophospho-phate by sequential addition of purine precursors that include glycine, amide groups from aspartate and glutamine, and methenyl and formyl tetrahydrofolates.
  • Synthesis occurs in the cytosol.

II. *Secondary Metabolism*

A. Terpenes and terpenoids
  • Classification of terpenes is based on a basic branched C_5 isoprene unit.
  • Isopentenyl diphosphate (IPP) is the fundamental precursor for terpenoids.
  • Plants emit about 15% of their fixed carbon into the atmosphere as isoprene.
  • Most terpenoids are produced, stored, and emitted in specialized structures, such as glandular trichomes, flower petals, and resin ducts.

# of C_5 units	Class Name	Examples
1	hemiterpenes[a]	isoprene
2	monoterpenes[a]	volatile essences of flowers, essential oils, pyrethrin
3	sesquiterpenes[b]	essential oils, abscisic acid
4	diterpenes[a]	gibberellins, phytoalexins, taxol, skolin
6	triterpenes[b]	sterols, brassinosteroids, oleanolic acid (surface waxes)
8	tetraterpenes[a]	carotenoids accessory pigments (photosynthesis)
> 8	polyterpenes[b]	plastoquinone, ubiquinone, dolichol, rubber
10	meroterpenes	partially derived from terpenoids; cytokinins, vincristine, vinblastine

[a] Synthesized in plastids.
[b] Synthesized in cytosol and endoplasmic reticulum.

B. Alkaloids
  • All alkaloids contain nitrogen, most are basic. Accumulate in actively growing tissue, epidermal and hypodermal cells, vascular sheaths and latex vessels; present in vacuoles.
  • Often stored in tissues other than where synthesized.
  • Formed from L-amino acids (tryptophan, tyrosine, phenylalanine, lysine, arginine) alone or with steroidal, secoiridoid, or other terpenoid-type moiety.
- *Aromatic amino acids.*
  • Phenylalanine and tyrosine give rise to peyote and morphine alkaloids, colchicines, and betalains.
  • Tryptophan gives rise to indole phytoalexins, indole glucosinolates, and indole alkaloids.

C. Phenylpropanoids
  • Derived from phenylpropanoid (C_6C_3) and phenylpropanoid-acetate ($C_6C_3\text{-}C_6$) skeletons.
  • Most phenolic compounds are derived from phenylpropanoids and include lignins, lignans and flavonoids.
  • Major classes of plant phenols are given below.

TABLE 9.1
— continued

# of C atoms	Class	Example	Source
6	phenols	catechol	*Galutheria* leaves
7	phenolic acids	ρ-hydroxybenzoic acid	widespread
8	phenylacetic acids	2-hydroxyphenylacetic acid	*Astilbe* leaves
9	hydroxycinnamic acids	caffeic acid	ubiquitous
	phenylpropenes	myristicin	*Myristica fragrans*
	coumarins	6-7,dimethoxycoumarin	*Dendrobium densiflorum*
		8-methoxypsoralen	*Heracleum mantegazzianum*
	isocoumarins	hydrangenol	*Hydrangea macrophylla*
	chromones	eugenin	*Eugenia aromatica*
10	napthoquinones	juglone	*Juglans nigra*
13	xanthones	mangiferin	widespread
14	stilbenes	resveratol, lunularic acid	*Vitis vinifera,* liverworts
	anthraquinones	emodin	rhubarb
15	flavonoids	flavones, catechins, isoflavones	soybean, green tea
18	lignans	pinoresinol	conifers
	neolignans	eusiderin	Magnoliaceae
30	bioflavonoids	amentoflavone	gymnosperms

From Goodwin, T.W. and Mercer, E.I. 1983. *Int. Plant Biochem.*, 2nd ed., Pergamon Press, NY, 677 pp.

- *Benzopyranones*
  • A group of defense-related compounds that include coumarins, stilbenes, styrlpyrones, and arylpyrones.
  • Coumarins can cause internal bleeding, photophytodermatitus, and are used to treat skin disorders.
  • Stilbenes, styrlpyrones, and arylpyrones are derived from cinnamoyl-CoA and malonyl-CoA and flavonoids pathways.

(Adapted from Goodwin T.W. and Mercer, E.I. 1983. *Intr. Plant Biochem*, 2nd ed., Pergamon Press, NY, 677 pp.; Mohr, H. and Schopfer, P. 1995. *Plant Physiology*, 4th ed., Springer-Verlag, Berlin, Heidelberg, 629 pp., Buchanan, B.B., Gruissem, W., and Jones, R.L. (eds.) 2000. *Biochem, Molecular Biology of Plants*, Am. Soc. Plant Physiology, Rockville, MD, 1367 pp. With permission.)

Complete biochemical knowledge of the pathway is not necessary. Although it is often research in the model systems that provide the requisite knowledge to improve cultivated species, the vast majority of unique and undiscovered phytochemicals are produced in noncultivated plants. *Arabidopsis thaliana* (mouse-ear cress or thale cress) serves as the model plant for plant biologists, and gene sequences of certain enzyme classes can be identified from the *Arabidopsis* database, with the caveat that genes encoding enzymes in natural product pathways are not closely linked in plants (Dixon, 2001). Conversely, the pathways for vitamins are widely conserved by both eukaryotes and prokaryotes.

Many phytochemicals have antioxidant properties, including vitamins C and E, ß-carotene, a variety of carotenoids, and plant phenols. It is generally thought that by increasing the dietary intake of phytochemicals with antioxidant properties, aging effects on cells and diseases can be delayed. Commercial preparations often consist of a mixture of antioxidants. For example, Pycnogenol® is a mixture of phenolic and

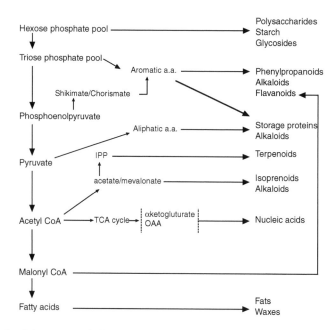

Figure 9.1 Primary metabolites, intermediate products and derived natural products. (Adapted from Goodwin, T.W. and Mercer, E.I. 1983. *Intr. Plant Biochem.*, 2nd ed., Pergamon Press, NY, 677 pp.; Mohr, H. and Schopfer, p. 1995. *Plant Physiology*, 4th ed., Springer-Verlag, Berlin, Heidelberg, 629 pp.; Buchanan, B.B., Gruissem, W., and Jones, R.L. (eds.) 2000. *Biochem, Molecular Biology of Plants*, Am. Soc. Plant Physiology, Rockville, MD, 1367 pp. With permission.)

polyphenolic compounds (Dillard and German, 2000). With mixtures, the specific bioactive chemical is usually not known, and it is possible the efficacy of the isolated compounds would diminish if ingested separately, i.e., mixtures of bioactive compounds may act synergistically. Improvement of the phytochemical content of plants using breeding or bioengineering requires that specific compounds be identified and quantified. Any step in the biosynthetic pathway could potentially be a target for manipulation, and it is theoretically possible that several intermediaries could be increased along the pathway in addition to the target metabolite.

THE ROLE OF PLANT BREEDING IN IMPROVING PHYTOCHEMICAL CONTENT

Traditionally, the primary goal of plant breeders has been to improve yield by developing varieties resistant to diseases, insects, and abiotic stress. Only recently have breeders attempted to change the phytochemical composition of plants for human health. The number of species that have had their nutritional content altered is few, and most of the attention within this area has been directed toward micronutrient (defined as most minerals and all vitamins) content, not phytochemicals

(Graham et al., 1999). The major emphasis has been on increasing the iron, zinc, and β–carotene content in wheat, maize, rice, beans, and cassava (Bouis, 1996). The FAO and WHO have identified micronutrient malnutrition as a primary health concern, affecting more than two billion people worldwide (Welch and Graham, 1999). In developed countries, however, caloric intake generally exceeds needs, and there has been an increased emphasis on improving micronutrient and phytochemical content of food.

Allard (1999) defined plant breeding as the "controlled evolution of plants ... by humans with the goal of producing populations that have superior agricultural and economic characteristics." Plant improvement by breeding is limited by naturally occurring variation and the restriction of reproductive barriers between distantly related plants. Although these conditions limit the breadth of genetic transfer that is possible, the relatively low cost of implementing and conducting a breeding program ensures that this method will continue to be used. Variation is key to plant improvement through breeding, and results from mutations are estimated to occur at a rate of 10^{-5} to 10^{-7}/locus/generation (Allard, 1999). Random mutations that affect amino acid sequences, and therefore protein function, normally have deleterious effects on plant survival. It is not likely, therefore, that either naturally occurring or chemically induced random mutations would generate useful plants with respect to phytochemical content. Further, identifying and characterizing the mutation is slow and is arguably the bottleneck of this technology, especially as it relates to developing useful screening procedures (Zhu, 2000). Recently developed methods such as T-DNA insertional mutagenesis offers an alternative to random mutations, and are used to introduce mutations and quickly characterize the function and location of the gene (Azpirox-Leehan and Feldmann, 1997; Krysan et al., 1999).

Most of our major, and many minor, domesticated crops are well represented in seed banks around the world. Within these collections are named cultivars and varieties, numbered accessions, and land races. A much smaller number of wild (nondomesticated) relatives are housed in seed banks, and these offer a source of rich genetic variation that has often been overlooked. Many of these species can be used as gene donors and crossed with the domesticated species. Potentially, there is great genetic diversity within a species having wide geographic distribution. Over time, in response to geographic and reproductive isolation, different multilocus assemblages of alleles would have accumulated in these populations in response to natural selection. It is likely that these accessions would differ significantly in many traits including vitamin, micronutrient, and phytochemical expression. Accessions from wild populations have commonly been used as sources for disease resistance (Tanksley and McCouch, 1997), and they should be screened for nutrient and phytochemical content. It is these traits in unimproved accessions and land races that breeders must identify and introgress into breeding lines. This task is a substantial undertaking because very little assessment of genetic diversity has been made of the germplasm held within gene banks.

BREEDING IS INHERENTLY INEFFICIENT

Breeding in its simplest form consists of the controlled transfer of pollen from one plant to another and subsequently selecting for the desired trait from among the progeny. Genetic recombination occurs during meiosis, when homologous chromatids exchange chromosomal segments, and again during chromosome assortment. Although crossover events are not random, the physical rearrangement of the chromosomes cannot be controlled *in situ*, and therefore the genetic composition of the gametes, and by extension, the phenotypes of the progeny, cannot be controlled. Specific single-gene transfer is not possible by plant breeding.

Frequently, unwanted genes are transferred along with the desired genes when they are in close physical proximity, a well-known phenomenon called genetic linkage. A process of repeated crosses between a recurrent parent and the progeny, followed by selection of the desired traits, must be performed to break the linkage and remove the unwanted genes. With many breeding schemes, selection must be performed with each generation and large numbers of plants must be grown to ensure that all genotypes are represented from the original cross. For example, in a cross between two parents that differ in only six loci, 4,096 F_2 plants must be grown and screened to recover each of the possible genotypes. In tomato, more than 20 loci have been found that affect carotenoid synthesis, so recovery of all the genotypes would theoretically require a very large population (Porter et al., 1984). Environmental variation from sources such as mineral composition of the soil, temperature, and light intensity and quality, must be minimized. Ideally, the most efficacious method in which to control these variables is to screen plants using hydroponics and a controlled environment chamber.

Selecting for phytochemical content also requires extraction and analysis of the phytochemical from tissue samples from every plant of the population. High throughput technology makes this feasible, but in many instances, greenhouses or suitable land and labor needed to grow large numbers of plants may not be available. It is well documented that metabolic pathways are controlled by multiple genes, and therefore, phytochemical content is influenced by the environment, genotype, tissue, and maturity of the plant. For example, Prior et al., (1998) found an increase in phenolics and anthocyanins of blueberries with plant maturity. Isoflavone content in soybean seeds was shown to vary with environment and variety (Eldridge and Kwolek, 1983; Tsukamoto, 1995). Breeders over the past 50 years have focused mainly on improving protein quantity and quality of staple foods, so information is largely lacking on the effects that agronomic practices, climate, and soil have on phytochemical levels. This area of research should be given high priority, as manipulating the cultural environment may be the easiest way to increase the phytochemical content of a plant (Graham et al., 1999; Scott et al., 2000).

Metabolic pathways are often controlled by multiple genes and complex feedback mechanisms, and therefore the need to identify genotype-by-environment (G:E) interaction is of paramount importance. Thus, plants must be grown in multiple environments for evaluating G:E interaction, which adds to the expense of breeding for improved phytochemical content. The efficiency and capabilities of plant breeding have been improved with the development of molecular markers to identify specific traits and the ability to treat quantitative traits as discrete Mendelian units (Lander and Botstein, 1989; Paterson et al., 1989). Once identified, these quantitative trait loci (QTL) can be mapped to chromosomal regions in much the same manner as single gene traits.

MARKER-ASSISTED SELECTION

Genetic recombination is a result of crossing-over between homologous chromosomes, and from the independent assortment of maternal and paternal chromosomes during meiosis. One cannot make predictions *a priori* where crossing-over will occur and what the genotype of the resulting progeny will be. Only by making crosses and developing segregating populations can linkage maps be constructed, and predictions made about the transmission and inheritance of genes. Marker-assisted selection (MAS) is a method in which DNA markers are associated with a specific trait or QTL. Markers are generated by crossing genetically and phenotypically dissimilar parents, evaluating the progeny of a segregating population from that mating, and associating a polymorphic marker with the trait. To be useful, the marker must be in close physical proximity to the gene of interest and, ideally, the marker will bracket the gene. Selection efficiency is increased because the DNA from each individual of the population can be assayed for the marker. Based on the presence or absence of the marker, the individual is either included or excluded from the gene pool. MAS allows culling of genetic material at a much earlier stage and linkage drag is minimized because a clearer relationship between genotype and phenotype is established, thereby increasing efficiency by lowering the number of plants required in each filial population. MAS is especially useful in backcross breeding in which single genes or QTLs are introgressed into elite germplasm that lack the trait of interest (Bernacchi et al., 1998; Tanksley and Nelson, 1996).

A distinct advantage of both conventional plant breeding and marker-assisted selection is that metabolic pathways, enzyme structures, or catalytic mechanisms do not need to be well defined (Schmidt-Dannert et al., 2000). It has been a long-held presumption, with no supporting evidence, that increasing the nutrient content might lower crop yield (Ruel and Bouis, 1998). It should be feasible to alter phytochemical or nutrient content without altering yield, because phytochemicals make up a very small percentage of a plant's dry weight (<0.1%) (DellaPenna, 1999). Most species show two- to three-fold differences of phytochemicals and antioxidants (Ruel and Bouis, 1998; Prior et al., 1998; USDA source of phytochemicals, 1998), yet no significant yield differences have been reported. Further, vitamin and micronutrient concentration can be increased severalfold without detriment to human health, with the notable exception of the toxic heavy metals Zn, Cu, and Ni (DellaPenna, 1999). Because of the expense involved with breeding for phytochemicals, more scientifically based data are needed which clearly define the relationship between phytochemicals and human health.

GENOMIC AND PROTEOMIC APPROACHES

Arabidopsis thaliana was the first higher plant to be sequenced (The *Arabidopsis* Genome Initiative, 2000). The same benefits to human health attributed to the human genome project will be realized for plants. Namely, the number of genes, their location, and most importantly, the ability to assign function to genes based upon the translation of the nucleotide sequence into amino acids and proteins. The *Arabidopsis* genome has been estimated to contain about 25,500 genes encoding 11,000 protein families (The *Arabidopsis* Genome Initiative, 2000). Thirty-five percent of the predicted proteins are unique to the *Arabidopsis* genome. Functional analysis of the proteins was based upon sequence similarity to proteins of known function of all organisms, which has been shown to share a surprising amount of trans-kingdom sequence homology at the primary protein level (DellaPenna, 1999). It is expected that many phytochemicals will be unique to one or a few species; these enzymes will not have homologs in *Arabidopsis*.

Nevertheless, Arabidopsis has been useful in elucidating phytochemical pathways using a variety of strategies. For example, Borevitz et al. (2000) using activation-tagged *Arabidopsis* lines found that different transcription factors were involved in the phenylpropanoid pathway, and the authors discussed how phytochemical production could be increased by activation tagging methodology. Taking advantage of the public database, a genomic approach based on fungal and human orthologs and *Arabidopsis* sequence data was used to increase α-tocopherol in seeds by 80-fold by over-expressing γ–tocopherol methyltransferase (Shintani and DellaPenna, 1998).

Proteomics has been defined as the systematic analysis of expressed proteins of a given genome (Jacobs et al., 2000). Because sequencing and transcription profiling does not directly give information about gene function, proponents in the field of proteomics believe a more efficient method of determining gene function can be realized by global protein profiling. The most complete protein database has been derived from yeast, where 6100 of the 6800 proteins have been identified and 56% of them experimentally characterized (Jacobs et al., 2000). In contrast, only 9% of the expressed proteins have been characterized experimentally in *Arabidopsis* (The *Arabidopsis* Genome Initiative, 2000). The question that remains to be answered is how much of the information gained from other (trans-kingdom) organisms will be useful in manipulating phytochemical content of plants. In terms of improving vitamin and micronutrient content, this information will be useful as the genes utilized in these pathways are highly conserved among plants as well as bacteria and yeast. Genetic databases can be used to identify orthologous genes, elucidate pathways, and serve as a blueprint to add enzymes to pathways in staple crops in which these are lacking (Croteau et al., 2000; DellaPenna, 1999; Ye et al., 2000). The

Arabidopsis databases contain a wealth of gene sequence information and it should be utilized to the extent possible, but many phytochemicals are limited to specific plant taxa, and therefore orthologs will not exist in *Arabidopsis* databases. Gene function is not known for approximately 30% of the *Arabidopsis* proteins and will have to be empirically determined. Thus, at this point, having the genomic sequence of a higher plant will not aid phytochemical engineering of pathways of rare secondary metabolites.

BIOENGINEERING

Bioengineering can be defined as the alteration of the gene expression of any biological organism by molecular methodologies. Bioengineering in any metabolic pathway requires that all steps within the pathway be defined, the regulatory points identified, and the stability of the intermediates determined (Scott et al., 2000). Phytochemicals are often differentially expressed both spatially and temporally. For example, the concentration of alkamides differs among plant organs in *Echinacea purpurea* with the roots having the highest concentration (Perry et al., 1997). To increase harvest efficiency, the production of any phytochemical should be directed to a readily harvestable organ, such as a leaf or seed. Using recombinant technology, expression cassettes can be made to redirect synthesis into different organs (Ye et al., 2000).

Elucidation of metabolic pathways can be accomplished by performing isotopic tracer experiments, screening for mutants, or gene silencing/overexpression of the putative genes in transgeneic plants. Modeling metabolic flux in combination with *in vivo* analysis must be performed to find control mechanisms. Earlier strategies worked under the premise that there is a rate-limiting enzyme, but there is general agreement that any enzyme in a pathway may affect flux. There are homeostatic genetic and metabolic control mechanisms that keep each enzyme at ideal levels (Jensen and Hammer, 1998); however, over-expression of a single enzyme, γ-TMT, increased tocopherol by 80-fold in *Arabidopsis* seeds, a clear indication that this was the rate-limiting enzyme (Shintani and DellaPenna, 1998).

Metabolic pathways from different organisms can be combined and expressed in *E. coli* and other organisms (Schmidt-Dannert et al., 2000). A well-reported success of this strategy was the expression of β–carotene in the endosperm of rice (Ye et al., 2000 and references within). β–Carotene is produced in plastids of plant cells by the isoprenoid pathway (Figure 9.2). The genes encoding phytoene synthase, δ–carotene desaturase, and lycopene β-cyclase were introduced into the rice endosperm to enable β–carotene production. The genes for phytoene synthase were taken from daffodil (*Narcissus pseudonarcissus*), while the latter two genes originated in *Erwinia uredovora*. These carotene pathway genes were placed into three vectors (Figure 9.3) and placed under control of an endosperm-specific (glutelin; Gt1 p) and a constitutive CaMV (cauliflower mosaic virus; 35S p) promoter. Correct expression of the genes also required that functional transit peptides for import into plastids be part of the construct (*crtI*, and *tp*, Figure 9.3). Transferring the pathway and direct-

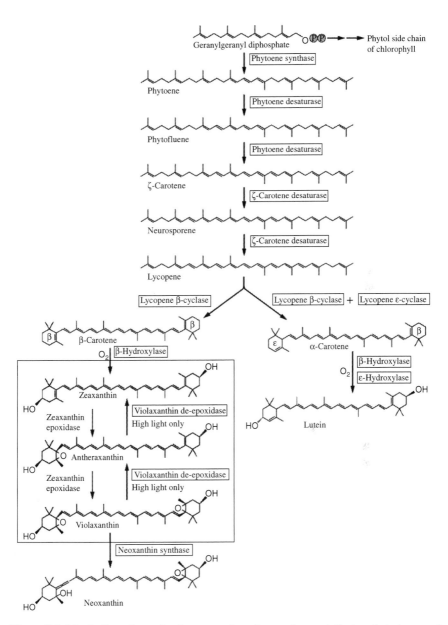

Figure 9.2 Metabolic pathway for the conversion of geranylgeranyl disphosphate to α- and β-carotene. Phenylpropanoid pathways have several control points from which many different natural products are synthesized. The metabolic flux between each control point, and within each pathway, may differ with respect to genetic composition of the species, cultivar or variety, the environmental conditions, tissue, and maturity of the plant. (From Buchanan, B.B., Gruissem, W., and Jones, R.L. (eds.) 2000. *Biochem, Molecular Biology of Plants*, Am. Soc. Plant Physiology, Rockville, MD, 1367 pp. With permission.)

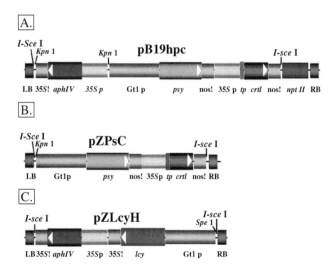

Figure 9.3 Vectors used to express phytoene synthase (*psy*), phytoene desaturase (*crtI*), and lycopene-β cyclase (*lcy*) for synthesis of β-carotene in rice endosperm (from Ye et al., 2000). A, Vector B19hpc was engineered to produce lycopene, and was introduced as a single transformant with *Agrobacterium*. B., C. Vectors pZPsC and pZLcyH were engineered to synthesize β-carotene, and were introduced as cotransformants with *Agrobacterium*. For proper spatial and temporal expression of the genes, the vectors included both a constitutive promoter (35Sp), and a glutelin promoter (Gtlp) and a transit peptide sequence (*tp*) for import and expression in endosperm plastids. For additional details concerning vector construction, the reader is referred to Burkhardt et al. (1997) and Ye et al. (2000). (From Ye, X., Al-Babili, S., Klöti, S., Zhang, J., Lucca, P., Beyer, P., and Poltrykus, I. 2000. Engineering the provitamin A (β-carotene) biosynthetic pathway into (carotenoid-free) rice endosperm. *Science*, 287:304. With permisssion.)

ing its expression to the endosperm was technically demanding and expensive, reportedly requiring $2.6 million and 7 years (Nash and Robinson, 2000), yet this example demonstrates the potential for interkingdom transfer and expression of metabolic pathways into plants.

EXPANDING POSSIBILITIES

Until now researchers in the field of bioengineering and plant breeding have altered the expression of existing enzymes or have introduced the expression of a phytochemical. These approaches have in common the fact that no novel phytochemicals are synthesized; no "new to nature" products have been made. New theories and technologies are being developed that are directed toward changing the efficiency of enzymes, and in these experiments, phytochemicals have been produced in microorganisms that are "essentially inaccessible from natural sources or by synthetic chemistry" (Schmidt-Dannert et al., 2000).

This recent body of work applies the principles of Darwinian evolution to alter enzyme efficiency and function (Arnold and Volkov, 1999; Stemmer, 1994a,b). Two separate strategies have been utilized; a rational design approach involving iterative computer protein design and site-directed mutagenesis of single nucleotides, and a second strategy in which blocks of nucleotides are shuffled to produce novel peptides, a method called "DNA shuffling" (Stemmer, 1994,a,b). The former method can be viewed as *in vitro* molecular evolution because *in vitro* recombination can occur between the DNA templates of two related genes that differ in their sequence, and the recombinants screened for improved performance (Stemmer, 1994,a,b). The DNA templates can contain random point mutations, single genes, or homologous genes (Crameri et al., 1998).

As mentioned above, a frequently used method in plant breeding is the backcross, in which a specific trait is introduced into a cultivar or variety and the desired genetic background is recovered by repeated crosses of the progeny to the recurrent parent. Similarly, by molecular backcrossing against the wild-type DNA, nonessential mutations can be selected against in the resulting recombinants (Stemmer, 1994). This relatively simple technique resulted in a 32,000-fold increase in antibiotic resistance (Stemmer, 1994a), and a 1,000-fold increase in substrate specificity of β-fucosidase (Zhang et al., 1997). Related sequences have been bred (Schmidt-Dannert et al., 2000), as well as sequences with only 50% homology (Ostermeier et al., 1999). One can imagine a strategy to improve metabolic processes in plants using *in vitro* molecular evolution.

To increase the efficiency with which candidate plants and phytochemicals are discovered, more sophisticated networks are being developed between ethnobotanists, medical professionals, and the agriculture/botany community. As the process matures, increasingly sophisticated experiments can be designed to more clearly define the role of phytochemicals in human health and plant function. Cell lines from specific tissues and defined genotypes can be used to screen the efficacy of phytochemicals. Genetic and metabolic profiling of plants will enable phytochemicals to be precisely identified and determine under what conditions they are produced. Information from cell screening can be used to direct the development of plants with unique phytochemical profiles. Any promising plant collected from the wild must be adapted to agriculture, and its propagation methods worked out. Most perennial, and many annual plants have seed dormancy mechanisms that make efficient agronomic production all but impossible. Further, the seed of many subtropical and tropical plants are recalcitrant, meaning they cannot withstand desiccation and therefore cannot be stored for appreciable lengths of time. Alternatively, whole metabolic pathways, or portions thereof, could be transferred into domesticated plants exhibiting good agronomic characteristics such as high biomass production and no seed dormancy.

ACKNOWLEDGMENTS

This work was supported by grants from the NIH: P50 AT00151 and the California State Agricultural Research Initiative.

REFERENCES

Allard, R.W. 1999. *Principles of Plant Breeding*, John Wiley & Sons, Inc., N.Y., 254 pp.

Arnold, F.H. and Volkov, A.A. 1999. Directed evolution of biocatalysts, *Curr. Op. Chem. Biol.* 3:54–59.

Azpiroz-Leehan, R. and Feldmann, K.A. 1997. T-DNA insertion mutagenesis in Arabidopsis: going back and forth, *Trends Genet.*, 13:152–156.

Bernacchi, D., Beck-Bunn, T., Eshed, Y., Lopez, J., Petiard, V., Uhlig, J., Zamir, D., and Tanksley, S. 1998. Advanced backcross QTL analysis in tomato. I. Identification of QTLs for traits of agronomic importance from *Lycopersicon hirsutum, Theor. Appl. Genet.*, 97:381–397.

Borevitz, J.O., Xia, Y., Blount, J., Dixon, R.A., and Lamb, C. 2000. Activation tagging identifies a conserved MYB regulator of phenylpropanoid biosynthesis, *Plant Cell,* 12:2383–2393.

Bouis, H. 1996. Enrichment of food staples through plant breeding: A new strategy for fighting micronutrient malnutrition, *Nutr. Rev.*, 54:131–137.

Buchanan, B.B., Gruissem, W., and Jones, R.L. (Eds.) 2000. *Biochem. Molecular Biology of Plants,* Am. Soc. Plant Physiol., Rockville, MD, 1367 pp.

Burkhardt, P.K., Beyer, P., and Wunn. J. 1997. Transgenic rice (*Oryza sativa*) endosperm expressing daffodil (*Narcissus pseudonarcissus*) phytoene synthase accumulates phytoene, a key intermediate of provitamin A biosynthesis, *Plant J.*, 11:1071–1078.

Crameri, A., Raillard, S.-A., Bermudez, E., and Stemmer, W.P.C. 1998. DNA shuffling of a family of genes from diverse species accelerates directed evolution, *Nature*, 391:288–291.

Croteau, R., Kutchan, T.M., and Lewis, N.G. 2000. Natural products (secondary metabolites), in *Biochem. Molecular Biology of Plants*, Buchanan, B., Gruissem, W., and Jones, R.L. (Eds.) Am. Soc. Plant Physiol., Rockville, MD, pp. 1250–1318.

DellaPenna, D. 1999. Nutritional Genomics: Manipulating plant micronutrients to improve human health, *Science*, 285:375–379.

Dillard, C.J. and German, J.B. 2000. Phytochemicals: nutraceuticals and human health, *J. Science Food Agric.*, 80:1744–1756.

Dixon, R.A. 2001. Natural products and plant disease resistance, *Nature,* 411:843–847.

Dudal, R. 1976. Inventory of the major soils of the world with special reference to mineral stress hazards. in *Plant Adaptation to Mineral Stress in Problem Soils*, Wright, M.J. (ed.), Cornell University Agricultural Experiment Station, Ithaca, NY, pp. 3–14.

Duke, S.O., Rimando, A.M., Dayan, F.E., Canel, C., Wedge, D.E., Tellez, M.R., Schrader, K.K., Weston, L.A., Smillie, T.J., Paul, R.N., and Duke, M.V. 2000. Strategies for the discovery of bioactive phytochemicals, in *Phytochemicals as Bioactive Agents*, Bidlack, W.R., Omaye, S.T., Meskin, M.S., and Topham, D.K.W. (Eds.), Technomic Publ. Co., Inc. Lancaster, PA, 274 pp.

Eldridge, A.C. and Kwolek, W.F. 1983. Soybean isoflavones: effect of environment and variety on composition, *J. Agric. Food Chem.*, 31:394–396.

Goodwin, T.W. and Mercer, E.I. 1983. *Intr. Plant Biochem.*, 2nd ed., Pergamon Press, NY, 677 pp.

Graham, R., Senadhira, D., Beebe, S., Iglesias, C., and Monasterio, I. 1999. Breeding for micronutrient density in edible portions of staple food crops: conventional approaches, *Field Crops Research*, 60:57–80.

Jacobs, D.I., van der Heijden, R., and Verpoorte, R. 2000. Proteomics in plant biotechnology and secondary metabolism research, *Phytochem. Anal.,* 11:277–287.

Jensen, P.R. and Hammer, K. 1998. Artificial promoters for metabolic optimization, *Biotech. Bioeng.,* 58:191–195.

Kramer, P.J. and Boyer, J.S. 1995. *Water Relations of Plants and Soils*, Academic Press, NY, 495 pp.

Krysan, P.J., Young, J.C., and Sussman, M.R. 1999. T-DNA as an insertional mutagen in Arabidopsis, *Plant Cell*, 11:2283–2290.

Lander, E.S. and Botstein, D. 1989. Mapping Mendelian factors underlying quantitative traits using RFLP linkage maps, *Genetics,* 121:185–199.

Mann, C.C. 1999. Crop scientists seek a new revolution, *Science*, 283:310–316.

Mohr, H. and Schopfer, P. 1995. *Plant Physiology*, 4th Ed., Springer-Verlag, Berlin, Heidelberg, 629 pp.

Nash, J.M. and Robinson, S. 2000. Grains of hope, *Time*, 156(5):38–46.

Ostermeier, M., Shim, J.H., and Benkovic, S.J. 1999. A combinatorial approach to hybrid enzymes independent of DNA homology, *Nat. Biotech.,* 17:1205–1209.

Paré, P.W. and Tumlinson, J.H. 1997. Induced synthesis of plant volatiles, *Nature,* 385:30–31.

Paterson, A.H., Lander, E.S., Hewitt, J.D., Peterson, S., Lincoln, S.E., and Tanksley, S.D. 1989. Resolution of quantitative traits into Mendelian factors by using a complete linkage map of restriction fragment length polymorphisms, *Nature*, 335:721–726.

Perry, N.B., van Klink, J.W., Burgess, E.J., and Parmenter, G.A. 1997. Alkamide levels in *Echinacea purpurea*: A rapid analytical method revealing differences among roots, rhizomes, stems, leaves and flowers, *Planta Medica*, 63:58–62.

Pinstrup-Andersen, P., Pandya-Lorch, R., and Rosegrant, M.W. 1999. World food prospects: critical issues for the twenty-first century, Int. Food Policy Res., Inst., Washington, D.C., 32 pp.

Porter, J.W., Spurgeon, S.L., and Sathyamoorthy, N. 1984. Biosynthesis in carotenoids, in *Isopentenoids in Plants:Biochemistry and Functions*, Marcel Dekker, NY, pp. 181–183.

Prior, R.L., Cao, G., Martin, A., Sofic, E., McEwen, J., O'Brien, C., Lischner, N., Ehlenfeldt, M., Kalt, W., Krewer, G., and Mainland, C.M. 1998. Antioxidant capacity as influenced by total phenolic and anthocyanin content, maturity and variety of *Vaccinium* species, *J. Agr. Food Chem.,* 46:2686–2693.

Ruel, M.R. and Bouis, H.E. 1998. Plant breeding: a long-term strategy for the control of zinc deficiency in vulnerable populations, *Am. J. Clin. Nutr.,* 68S:488S–494S.

Scott, J., Rebeille, F., and Fletcher, J. 2000. Folic acid and folates: the feasibility for nutritional enhancement in plant foods, *J. Sci. Food Agric.,* 80:795–824.

Schmidt-Dannert, C., Umeno, D., and Arnold, F.H. 2000. Molecular breeding of carotenoid biosynthetic pathways, *Nature Biotech.,* 18: 750–753.

Stemmer, W.P.C 1994a. Rapid evolution of a protein *in vitro* by DNA shuffling, *Nature*, 370:389-391.

Stemmer, W.P.C. 1994b. DNA shuffling by random fragmentation and reassembly: *In vitro* recombination for molecular evolution. *Proc. Natl. Acad. Sci. USA*, 91:10747-10751.

Shintani, D. and DellaPenna, D. 1998. Elevating the vitamin E content of plants through metabolic engineering, *Science*, 282:2098–2100.

Tanksley, S.D. and McCouch, S.R. 1997. Seed banks and molecular maps: Unlocking genetic potential from the wild, *Science*, 277:1063–1066.

Tanksley, S.D. and Nelson, J.D. 1996. Advanced backcross QTL analysis: a method for the simultaneous discovery and transfer of valuable QTLs from unadapted germplasm into elite breeding lines, *Theor. Appl. Genet.*, 92:191–203.

The *Arabidopsis* Genome Initiative. 2000. Analysis of the genome sequence of the flowering plant *Arabidopsis thaliana, Nature*, 408:796–815.

Tsukamoto, C. 1995. Factors affecting isoflavone content in soybeans seeds: changes in isoflavones, saponins and composition of fatty acids at different temperatures during seed development, *J. Agric. Food. Chem.*, 43:1184–1192.

Welch, R.M. and Graham, R.D. 1999. A new paradigm for world agriculture: meeting human needs. Productive, sustainable, nutritious, *Field Crops Res.*, 60:1–10.

Ye, X., Al-Babili, S., Klöti, A., Zhang, J., Lucca, P., Beyer P., and Potrykus, I. 2000. Engineering the provitamin A (β-carotene) biosynthetic pathway into (carotenoid-free) rice endosperm. *Science*, 287:303-305.

Zhang, J.-H., Dawes, G., and Stemmer, W.P.C. 1997. Directed evolution of a fucosidase from a galactosidase by DNA shuffling and screening, *Proc. Natl. Acad. Sci. U.S.A.*, 94:4504–4509.

Zhu, J.-K. 2000. Genetic analysis of plant salt tolerance using *Arabidopsis, Plant Physiol.*, 124: 941–948.

CHAPTER 10

Phytochemicals in Oilseeds

FEREIDOON SHAHIDI

CONTENTS

1-58716-083-8/02/$0.00 + $1.50
© 2002 by CRC Press LLC

INTRODUCTION

Oilseeds serve as a main source of edible vegetable oils, including essential fatty acids belonging to both the ω3 and ω6 families, but also include a considerable amount of protein and carbohydrate. Furthermore, oilseeds contain minor components in both their oil and deoiled meal. The minor components in oleaginous seeds belong to a wide array of compounds representing different classes of phytochemicals which are often bioactive; these include phenolics and polyphenolics, tocopherols and tocotrienols, glucosinolates, phytates, phytosterols, lignans, carotenoids, lectins, and enzyme inhibitors, among others. The type of phytochemicals and their contents in different oilseeds depend primarily on the species concerned as well as agronomic, seasonal, and environmental conditions. This chapter provides a brief discussion of oilseed phytochemicals with specific emphasis on their phenolic and polyphenolic constituents.

PHENOLICS AND POLYPHENOLICS

BIOSYNTHESIS AND HEALTH EFFECTS

Phenolic compounds are the most widely distributed secondary metabolites in plants, and constitute several thousands of compounds. Phenolics in plants are primarily responsible for their protection from free radical stress under photosynthetic conditions and ultraviolet light, and act against herbivores and pathogens (Shahidi, 2000a). They also contribute to the variety of color and taste of foods containing them (Shahidi and Naczk, 1995). In addition, they serve as wound-healing agents in plants as a result of their oxidation and subsequent condensation with free amino acids and proteins.

Structurally, phenolics are derived, as a first step, from phenylalanine and in a small number of plants from tyrosine via the action of ammonia lyase. These compounds generally contain at least one aromatic ring with one hydroxyl group (phenols) or more (polyphenols). In oilseeds, they exist as low-molecular-weight compounds which occur universally in higher plants with only some species specificity, and oligomeric and polymeric forms. The resultant compounds from the action of ammonia lyase on aromatic amino acids, known as phenylpropanoids, may then be subjected to a variety of modifications in plants, including hydroxylation and methylation, to afford a wide range of C_6 - C_3 compounds which are derivatives of *trans*-cinnamic acid (Figure 10.1). These compounds may lose a two-carbon moiety to yield benzoic acid derivatives. Condensation of C_6 - C_3 compounds with malonyl coenzyme A affords chalcones which may subsequently cyclize, under acidic conditions, to produce flavonoids and isoflavonoids as well as related compounds with C_6 - C_3 - C_6 units (Figure 10.2), among others (Shahidi, 2000b).

Phenolic acids, phenylpropanoids, and flavonoids/isoflavonoids in foods may occur in the free form, but are often glycosylated with sugars, especially glucose. While the presence of sugars in such compounds is responsible for their specific

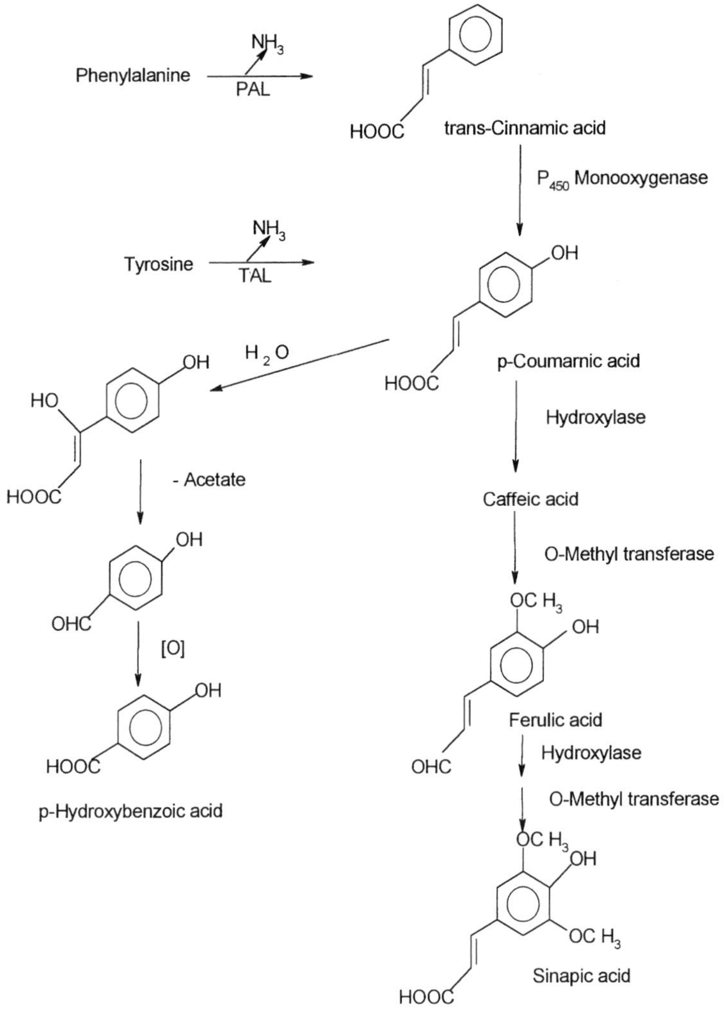

Figure 10.1 Biosynthesis of phenylpropanoids (C_6 - C_3) and benzoic acid derivatives from phenylalanine and tyrosine. PAL, phenylalanine ammonia lyase; TAL, tyrosine ammonia lyase.

characteristics and transport in the plants and/or body fluids, they do not have any significant effect on the biological activity of compounds involved once ingested. Nonetheless, when measuring total antioxidant capacity of oilseeds and their extracts, it might be necessary to hydrolyze them to free their phenolic hydroxyl groups that are responsible for their antioxidant behavior *in vivo*. Phenolic acids may also be present in the esterified as well as bound forms (Naczk and Shahidi, 1989).

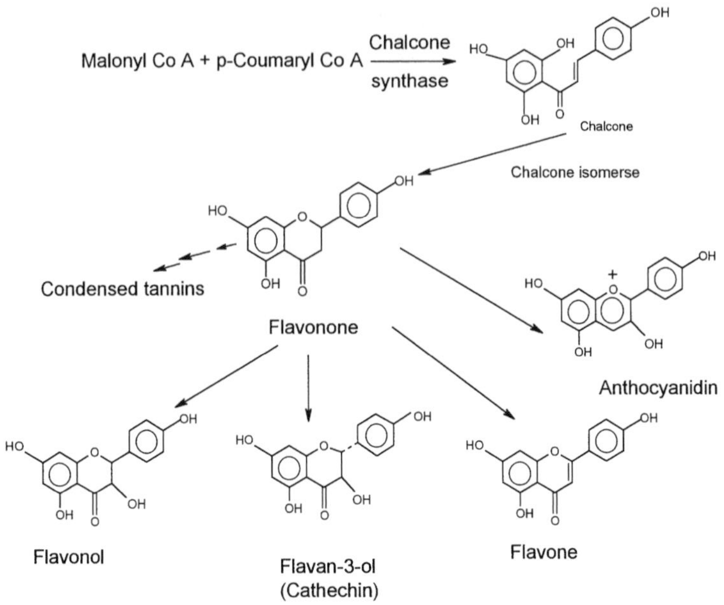

Figure 10.2 Biosynthesis of flavonoids (C_6 - C_3 - C_6) and condensed tannins.

Oilseed phenolics may exist in both simple and complex forms; the latter group consists of both hydrolyzable and condensed tannins. Condensed tannins are produced via polymerization of flavonoids, and are abundantly present in woody plants and seed coats, but are distinctly absent in herbal species. However, hydrolyzable tannins are formed by the reaction of gallic acid with hexose molecules and are more selectively present in 15 out of the 40 orders of dicotyledons. Thus, oilseeds contain a cocktail of different phenolics that may act cooperatively and synergistically with one another to exert their effects, both in terms of antioxidative action (Shukla et al., 1997) and health promotion and disease prevention.

In the body, oxidation products and reactive oxygen species (ROS) may lead to a number of diseases and tissue injuries such as those of the lungs, heart, kidney, liver, gastrointestinal tract, blood, eye, skin, muscle, and brain, as well as the aging process. In healthy individuals, ROS are neutralized by the action of antioxidant and antioxidant enzymes. However, when the action of the enzyme system is inadequate because of illness and during infancy or due to aging, the oxidation process is not controlled naturally, and augmentation may provide the necessary means to combat degenerative diseases and other ailments caused by ROS. The manner in which antioxidants intervene is by their effect in a multistage process and may involve prevention of lipid oxidation, protein cross-linking, and DNA mutation, among others (Shahidi, 1997).

In terms of neutralizing free radicals, phenolics are well known to protect cells and their components against cancer development. Simple phenolic acids and tocopherols have been shown to be potent inhibitors of formation of carcinogens such as N-nitroso compounds (Kuenzig et al., 1984). Meanwhile, inhibition of benzo (a) pyrene-induced neoplasia in the forestomach of mice fed various plant phenolics has been reported by Wattenberg et al. (1980). Chromosomal aberrations induced by polycyclic aromatic hydrocarbons were inhibited by caffeic acid (Raj et al., 1983), while chlorogenic acid blocked chemically induced carcinogens in the large intestine of hamsters (Mori et al., 1986). Chang et al. (1985) have demonstrated antitumor-promoting activity of ellagic acid and quercetin. Flavonoids, including catechins, were also found to reduce hyperlipedemia in animals (Choi et al., 1991). In the case of heart disease, inhibition of LDL cholesterol oxidation helps in prevention of foam cells formation and lipid streaks development. Oxidized LDL cholesterol is more atherogenic than native LDL and is also known to affect tissue factor expression. Several recent studies provide compelling evidence that dietary intake of antioxidants can lower the production of atherogenic oxidized LDL cholesterol and thus may decrease the risk of cardiovascular disease (Steinberg et al., 1989; Niki, 1991, Niki et al., 1995).

OCCURRENCE AND CONTENTS

The occurrence, content, and type of phenolics in oilseeds is dictated by the species involved; phenolic acids and phenylpropanoids are often prevalent in many oilseeds. Canola and other cruciferae seeds contain considerable amounts of phenolic compounds compared to other oilseeds. The total content of phenolic acids, including phenylpropanoids, in defatted rapeseed flour has been reported to be approximately 623–1281 mg/100g (Krygier et al., 1982; Kozlowska et al., 1983a,b; Naczk and Shahidi, 1989). Sinapic acid was recognized as a predominant phenolic compound of canola while *p*-hydroxybenzoic, vanillic, genistic, protocatechuic, syringic, *p*-coumaric, ferulic, and caffeic acids were among the minor phenolics present (Krygier et al.,1982; Dabrowski and Sosulski, 1984). Recently, Wanasundara et al. (1994) reported that the most active phenolic antioxidant in canola meal was 1-O-ß-D-glucopyranosyl sinapate. Quercetin and isorhamnetin were isolated in less than 1 ppm from canola/rapeseed, while rapeseed contained a number of condensed tannins composed of cyanidin, pelargonidin, kaempferol, and their derivatives. While canola hulls serve as a rich source of condensed tannins, cotyledons possess little tannins (Blair and Reichart, 1984; Naczk et al., 1994). The acetone extractable condensed tannins of rapeseed hulls were composed of leucocyanidin units (Leung et al., 1979). The hulls may be separated by dehulling and the isolated tannins may potentially be used as a source of raw material for functional food applications (Naczk et al., 2000).

Mustard seeds also contain a considerable amount of phenolic acids; the levels being similar, to those in rapeseed and canola. The most abundant phenolics in mustard were *trans*-sinapic and *p*-hydroxybenzoic acids. In addition, *cis*-sinapic and *trans*-ferulic acids were present in relatively high amounts (Dabrowski and Sosulski, 1984). The antioxidant activity of ethanolic extracts of mustard flour were attributed to the presence of polyhydroxyphenols, including flavones and flavonols (Shahidi et al., 1994).

The phenolics of importance to soybeans are the isoflavones, and their content in raw soy is 0.1–0.5%. However, during processing, a considerable loss of isoflavones may occur. Soybean germ is a rich source of isoflavones, 2.5–3.5%. These include genistein, daidzein, and glycitin, all present mainly as glycosides, and minor amounts as aglycones. A very small amount of coumesterol (6.05 ppm) was also present. Meanwhile, presence of prunetin, formononetin, and 4',6',7'-trihydroxy-isoflavone (Foote et al., 1970) has been reported in soybean. Soybean flour also contained chlorogenic, isochlorogenic, caffeic, ferulic, p-coumaric, syringic, vanillic, p-hydroxybenzoic, salicylic, and sinapic acids (White and Xing, 1997). It should be noted that the distribution of isoflavones in the germ is different from that of the whole seed. Thus, the biological activity of phytochemicals may depend on their distribution in different seed parts. Soybean isoflavones have been reported to have both estrogenic and antiestrogenic properties. Several epidemiological studies have shown excellent correlation between soy protein consumption and reduced risk of cancer as well as inhibition of tumor formation (Boyd, 2001). Other beneficial health effects have also been attributed to the phytochemicals present in soybean products (Hirota et al., 2000). Thus, expanding utilization of soybean and its products in foods has been encouraged (Wolf, 1983; Liu, 2000).

The presence of chlorogenic, quinic, and caffeic acids as major phenolics in sunflower meal has been reported; chlorogenic acid being most predominant at 2.7% (Cater et al., 1972). Quinic acid was present at 0.38% and caffeic acid at 0.2% (Cater et al., 1972). In addition, p-hydroxybenzoic, syringic, trans-p-coumaric, trans-ferulic, and vanillic acids were present in small amounts in sunflower flour (Dabrowski and Sosulski, 1984). Rutin and quercetin glycosides have been identified in cottonseed; however, the main phenolic in glanded cottonseed is gossypol which is present at 1.1–1.3% (Lawhon et al., 1977). This polyphenolic compound has antifertility properties. Cottonseed also contained trans-ferulic, trans-p-coumaric, trans-caffeic, and p-hydroxybenzic acids in minute amounts. Peanut extracts contained dihydroquercetin and taxifolin, as well as trans-p-coumaric acid and minor amounts of trans-ferulic, trans-sinapic, p-hydroxybenzoic, and trans-caffeic acids (Pratt and Miller, 1984).

The phenolic acid content of flaxseed was considerably lower than that of other oilseeds (Kozlowska et al., 1983a; Dabrowski and Sosulski, 1984; Wanasundara and Shahidi, 1994). Trans-ferulic acid was the predominant phenolic acid present, but trans-sinapic, trans-p-coumaric, trans-caffeic, and p-hydroxybenzoic acids were found in smaller amounts. In addition, presence of 0-coumaric, genistic, and vanillic acids was reported (Kozlowska et al., 1983a,b; Dabrowski and Sosulski, 1984). Flaxseed also contained a considerable amount of secoisolericresorsinol diglucoside (SDG; Amarowicz et al., 1994), a lignan with anticarcinogenic and chemopreventive activity (Westcott and Muir, 2000). Furthermore, the presence of matairesinol, another lignan, has been reported in smaller amounts in flaxseed (Adlercreutz and Mazur, 1997; Mazur and Adlercreutz, 1998), together with a number of polymeric phenolics (Bakke and Klostermann, 1956; Westcott and Muir, 1996, 2000). SDG and its corresponding aglycone are converted to enterodiol and enterolactone, once consumed. These latter metabolites have been postulated to possess antiestrogenic properties (Adlercreutz, 1997; Adlercreutz and Mazur, 1997).

The phenolic acids in sesame include *trans*-caffeic, *trans*-*p*-coumaric, and *trans*-ferulic acids in the decreasing order of abundance (Dabrowski and Sosulski, 1984). However, vanillic, syringic, sinapic, and 0-coumaric acids were also present (Kozlowska et al., 1983a). Two lignans, sesamin and sesamolin, were found in crude sesame oil at 0.44 and 0.25%, respectively. However, during the refining process these lignans may be converted to their corresponding alcohols. Thus, following bleaching of sesame oil, a drastic change in the amounts of sesamin, episesamin, and sesamolin occurred. The compounds sesamol, sesamolinol, sesaminol, pinoresinol, and a sesamol dimer were present in the processed oils (Fukuda et al., 1985). In addition, the sesame cake contained a number of glucosides of pinoresinol as well as those of sesaminol. These latter compounds, considered as pro-antioxidants, are converted to potent antioxidants once the meal is treated with ß-glucosidase to release the corresponding aglycones (Namiki, 1995).

Recently, phenolic components of borage and evening primrose meal were reported by Wettasinghe and Shahidi (1999a,b). Borage meal contained rosmarinic acid, syringic acid, and sinapic acid, while evening primrose contained gallic acid, (+) catechin and (-) epicatechin, and a high-molecular-weight polyphenol known as oenothin B (Shahidi, 2000a). The latter compound has also been isolated from leaves and branches of evening primrose as well as other plants (Hatano et al., 1990). This compound was found to serve as an effective antitumorigenic and anticarcinogenic agent (Miyamoto et al., 1993); thus its potential health-promoting activity requires further attention. The reactive oxygen species - and 2,2-diphenyl-1-picrylhydrazyl radical (DPPH$^\bullet$) - scavenging activity of borage and evening primrose extracts have been documented (Shahidi et al., 2000, Wettasinghe and Shahidi, 2000). Dietary evening primrose was also found to have a hypocholesterolemic effect in rats (Balasinska, 1998).

In general, the beneficial health effects of seed phenolics, in part, arise from their antioxidant activity. Although antioxidant activity may arise from contributions of different bioactives, phenolic compounds are among the most powerful antioxidants known in food sources. In general, the type of phenolics and their substitution patterns determine the antioxidant activity of molecules involved. Thus, structure-antioxidant activity relationship of phenolics in oilseeds is of interest. Presence of a second hydroxyl or a methoxy group in the ortho- or para-position of benzoic acid derivatives, phenylpropanoids, or flavonoid/isoflavones is known to enhance the antioxidant activity of compounds involved. While the presence of two hydroxyl groups in the ortho- or para-position may lead to production of stable quinoid-type structures, a second methoxy group in the ortho- or para-position is an effective electron donor that stabilizes the free radicals formed, thus enhancing even further the activity of compounds involved. Phenylpropanoids are more effective antioxidants when compared with their phenolic acid counterparts. This is due to further stabilization of the free radicals formed via the extended conjugation arising from the propene moiety of the molecules.

In flavonoids, structural parameters play a prominent role in the antioxidant efficacy of the compounds involved. Therefore, presence of an ortho-diphenolic group in the B ring, a 2-3 double bond conjugated with the 4-oxo function, and occurrence of hydroxyl groups in positions 3 and 5 are most important structural features dictating the efficacy of flavonoids as antioxidants (Bors et al., 1990; Ratty and Das, 1988).

TOCOPHEROLS, TOCOTRIENOLS, AND UBIQUINONES

Unsaponifiable matter present in the extracted oil from oilseeds contain a variety of components, some of which are volatile and may be partially stripped during the deodorization process. The content of tocols, referred to as the total amount of tocopherols (T) and tocotrienols (T3), varies with the type of oilseed; their distribution is also species-dependent. The members of each of the T and T3 families are designated as alpha, beta, gamma, and delta, depending on the number and position of methyl groups on the chroman ring (Figure 10.3) (Kamal-Eldin and Appelqvist, 1996). Furthermore, there are eight different isomers for each of the tocopherols (three chiral centers) and the vitamin E activity of each is based on that of the d-α-tocopoherol, which is an RRR isomer. Other isomers have activities that are always less than that of the natural isomer (Papas, 1999). With respect to their antioxidant activities *in vitro*, generally δ-and γ-tocopherols are most effective tocopherols, and tocotrienols are still more effective than tocopherols (Dziezak, 1986).

In oilseed lipids, there appears to be a direct relationship between the degree of unsaturation, as represented by the iodine value, and the total content of tocopherols (Shahidi and Shukla, 1996). Most vegetable oils contain an abundance of alpha-, gamma-, and delta-tocopherols (Table 10.1), while the beta-isomer is less prevalent, except in wheat germ oil. Meanwhile, tocotrienols are present mainly in palm, rice bran, and wheat germ oils. In view of new developments in the genetic modification of oilseeds, the content and profile of minor components, including those of tocols, may vary considerably in different varieties. However, the importance of seasonal variation, geographical area of cultivation, and degree and type of processing cannot be ignored.

With respect to ubiquinones, also known as coenzyme Q, these compounds occur naturally as 6 to 10 isoprenoid units, designated as Q_6 (UQ-6) to Q_{10} (UQ-10). Coenzyme Q_{10} (UQ-10) and to a lesser extent UQ-9 (Figure 10.3) have been found in edible oils from vegetable origin (Table 10.1). Ubiquinol is the reduced form of ubiquinone and may protect cell membranes from oxidizing radicals by a direct radical-scavenging mechanism (Halliwell and Gutteridge, 1999). Ubiquinol provides efficient protection *in vivo* for mitochondria against oxidation, similar to vitamin E in the cell.

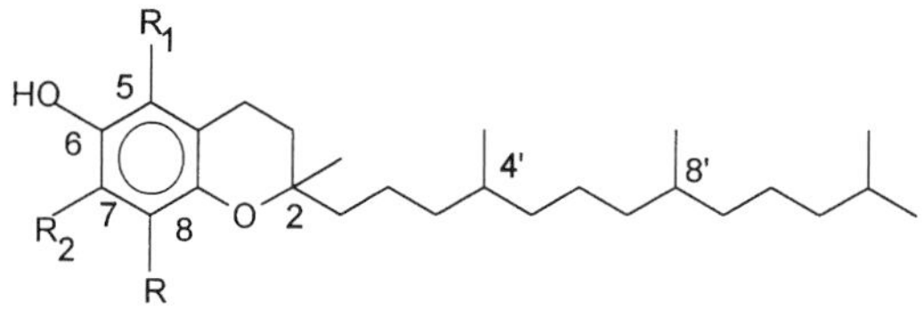

Tocopherol

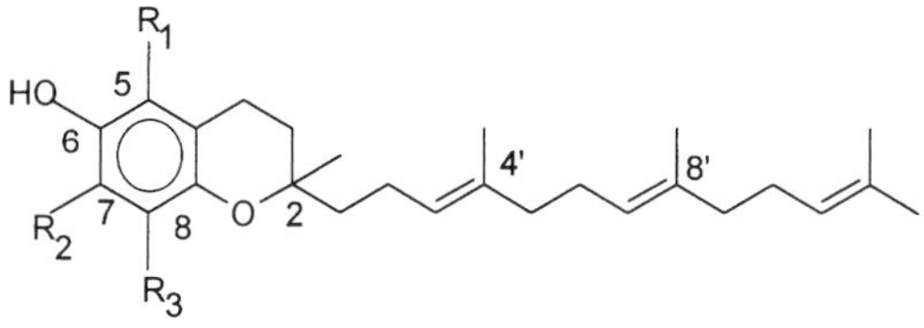

Tocotrienol

	R_1	R_2	R_3
α	CH3	CH$_3$	CH$_3$
β	CH$_3$	H	CH$_3$
γ	H	CH$_3$	CH$_3$
δ	H	H	CH$_3$

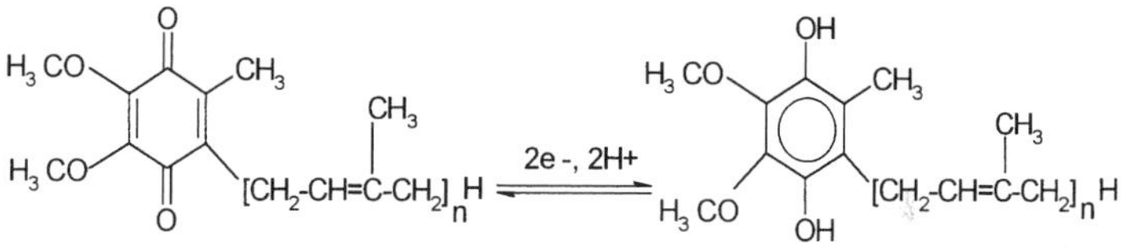

Oxidized Coenzyme Q   Reduced Coenzyme Q

Figure 10.3 Chemical structures of tocopherols, tocotrienols, and coenzyme Q's.

OTHER PHYTOCHEMICALS

A wide array of nonphenolic phytochemicals are present in oilseeds (Table 10.2). These include phospholipids, phytosterols, phytic acid, carotenoids, saponins, enzyme inhibitors, lectins and haemagglutinins, glucosinolates, oligosaccharides of sucrose, and cyanogenic glycosides, among others. The following sections provide a cursory account of these phytochemicals in oilseeds.

TABLE 10.1
Tocopoherols (T) and Ubiquinones (UQ) in edible oils (mg/100g)

Oil	α–T	β–T	γ–T	δ–T	UQ-9	UQ-10
Canola/rapeseed	17–26	—	35–61.2	0.4–1.2	0.2	7.3
Soybean	10.1–10.2	0.27	47.3–59.3	26.4–35.6	0.8	9.2
Sesame	1.0–1.8	0.1	24.4–51.2	3.2	—	3.2
Sunflower	48.7–78.3	0.25	1.9–5.1	0.7–1.0	2.1	0.4
Peanut	13.9–26.1	3.0–3.9	13.1–34.3	9.2–2.0	—	5.4
Cottonseed	38.9	—	38.7	—	ND	ND
Evening primrose	16	—	42	7	ND	ND

PHOSPHOLIPIDS

Phospholipids were, until recently, known to be responsible for off-flavor development in vegetable oils and other lipid-containing foods during storage. The oxidation sensitivity of phospholipids has been attributed to the higher degree of unsaturation of their fatty acid constituents. Phospholipids constitute the structural lipids of plant cell membranes and account for up to 10% of total lipids in oilseeds, depending on the type of seed examined (Belitz and Grosch, 1987). However, more recent studies have demonstrated that phospholipids may act as antioxidants, as they exert a synergistic effect when present together with tocols and ascorbates. Nonetheless, the exact mechanism and mode of action of phospholipids remains speculative. With respect to the claims related to the potential memory improvements upon consumption of phosphatidylserine, although the presence of docosahexaenoic acid (DHA) in brain phospholipids was originally thought to be responsible for the effects, more recent studies have provided evidence that the cognition-enhancing effects originally ascribed to bovine cortex were duplicated by phosphatidylserine from egg and soybean sources (Blokland et al., 1999). In addition, choline and phosphatidylcholine (lecithin) have been shown to have beneficial effects in enhancing cognition (Ladd et al., 1993; Meck and Williams, 1999).

PHYTOSTEROLS

Sterols constitute a major portion of unsaponifiable matter in oilseeds. These are composed of free sterols, sterol esters of fatty acids, and possibly sterol glucosides and acylated sterol glucosides (Shahidi and Shukla, 1996). Plant sterols are classified into three distinct groups belonging to the cholestane series, i.e., the phytosterol known as 4-desmethylsterols, 4-α-methylcholestane series known as 4-monomethylsterols, and lanostane series, that is triterpene alcohols known as 4,4'-dimethylsterols. Sterols are known to have antipolymerizing effect during frying (Sims et al., 1972; Yan and White, 1990).

The phytosterols are 28- and 29-carbon compounds including brassicasterol, campesterol, stigmasterol, ß-sitosterol, fucosterol, δ-avenasterol, and α-spinasterol.

TABLE 10.2
Distribution of various phytochemicals and/or their breakdown products in selected oilseeds

Oilseeds	Tocols	Other Phenolics	Lignans	Phospholipids	Phytosterols	Phytic acid	Carotenoids	Saponins	Enzyme Inhibitors	Glucosinolates	Oligsaccharides	Cyanogens
Canola	✓	✓	—	✓	✓	✓	✓	—	✓	✓	✓	—
Mustard	✓	✓	—	✓	✓	✓	✓	—	✓	✓	✓	—
Soybean	✓	✓	—	✓	✓	✓	✓	✓	✓	—	✓	—
Flax	✓	✓	✓	✓	✓	✓	✓	—	✓	—	✓	✓
Sesame	✓	✓	✓	✓	✓	✓	✓	—	✓	—	✓	—
Borage	✓	✓	—	✓	✓	✓	✓	—	✓	—	✓	—
Evening primrose	✓	✓	—	✓	✓	✓	✓	—	✓	—	✓	—

The major sterols of oilseeds are generally ß-sitosterol, stigmasterol, and campesterol. A majority of oils contain 100–500 mg sterols /100g sample. Notable exceptions are corn, rapeseed/canola, rice bran, and sesame oils, which contain up to 1.4, 1.0, 3.2, and 2.9 g sterols/100g sample, respectively. Rapeseed and mustard oils contain brassicasterol, which is characteristic of the cruciferae family and is present at 55–95 and 66–88 mg/100g oil, respectively (Shahidi and Shukla, 1996).

Phytosterols, at least some of them, are important inhibitors of cholesterol absorption by interfering with cholesterol synthesis and enhancing cholesterol excretion (Hicks and Moreau, 2001). They also exert anticancer properties and cytotoxicity (Piironen et al., 2000).

PHYTIC ACID

Phytic acid or *myo*-inositol 1,2,3,4,5,6-hexakis-dihydrogen phosphate is present at 1–6% in oilseeds and their defatted meals (Harland and Obertas, 1977), and serves as a source of storage phosphorus in plants (Rickard and Thompson, 1977). The phytic acid content, on a fat-free and dry basis, is 2–4% in canola and rapeseed, 2.4–2.7% in flax, 1.9% in peanut, 5.3% in sesame, 1.4% in soybean, and 1.9% in sunflower. Other oilseeds have similar levels of phytic acid. Of the 12 replaceable protons in phytic acid, 6 could be easily dissociated, thus giving rise to a highly negatively charged molecule which can bind ions of Ca, Mg, and Zn. By virtue of its ability to chelate iron, phytic acid is a potent inhibitor of iron-driven hydroxyl radical formation. Phytic acid also complexes with protein and starch molecules under physiological pH conditions, and hence may be regarded as an antinutrient. Nonetheless, recent evidences suggest that phytic acid has healthful properties. Reduction in glycemic response to starchy foods, as well as lowering of plasma cholesterol and triacylglycerol levels, have been observed with endogenous phytate consumed in foods. In addition, phytic acid has anticancer effects in the colon and mammary glands in rodent models and in various tumor cell lines (Rickard and Thompson, 1997). Furthermore, beneficial effects of phytic acid in relation to cardiovascular disease have been attributed to its hypocholesterolemic properties and chelation of metal ions that inhibits oxidation.

CAROTENOIDS

Carotenoids are widespread in oilseeds and particularly in all organisms that are exposed to light. In oilseeds, ß-carotene and other carotenoids are present in small amounts. However, these compounds are largely removed or decomposed during processing. Palm oil contains the highest content of carotenoids at a 500–700 ppm level (Goh et al., 1985). The beneficial health effects of carotenoids are attributed to their anticarcinogenic activity. Furthermore, the role of carotenoids as scavengers of singlet oxygen deserves attention. In this relation, carotenoids serve as the best suppressant in erythropietic protoporphyria, which results from the action of sunlight on individuals suffering from this condition.

SAPONINS

Saponins are made of a steroid (or triterpene) group attached to a sugar moiety. These surface-active compounds are found in legumes and oilseeds such as soybean. Saponins are able to lyse erythocytes due to their interaction with cholesterol in the membranes, thus exerting hemolytic properties. These compounds have a bitter taste and are toxic at high concentrations. Their surface activity is of interest due to their beneficial biological effects. The hypocholesterolemic effect of saponins is quite strong, especially when fed in the presence of cholesterol. Saponins also have anticarcinogenic properties and contribute to the enhancement of the immune system (Rao and Koratkar, 1997).

ENZYME INHIBITORS

Protease and amylase inhibitors are found in legumes and certain oilseeds (Whitaker, 1997). Protease inhibitors are easily denatured by heat, as are the amylase inhibitors. The antinutrient activity of protease inhibitors is due to their growth inhibitory effect and pancreatic hypertrophy. While potential health benefits of protease inhibitors remain unclear, lower incidences of pancreatic cancer have been observed in populations where the intake of soybean and its products is high.

Amylase inhibitors are known to have hypoglycemic effects, and trypsin inhibitors in soybean are known to have anticarcinogenic properties and are also effective in lowering caloric intake.

LECTINS AND HAEMAGGLUTININS

These are sugar-binding proteins that may bind and agglutinate red blood cells (Liener, 1997). They occur in most plant foods that are often consumed in the raw form. Although lectins in soybeans and peanuts are not toxic, those from many other beans, such as lima beans, are toxic when taken orally. The toxicity of lectins arises from their binding with the specific receptor sites on the epithelial cells of the intestinal mucosa with subsequent lesions and abnormal development of microvillae. The lectin-induced disruption of the intestinal muscosa may allow entrance of the bacteria and their endotoxins to the blood as well as other complications.

GLUCOSINOLATES

Brassica oilseeds contain a variety of glucosinolates that have traditionally been considered as antinutritional factors with goitrogenic effects (Shahidi et al., 1997). Their breakdown products, upon the action of myrosinase, are nitriles, thiocyanates, isothiocyanates, and possibly oxazolidinethione. Although glucosinolate degradation products are toxic, they are also responsible for the desirable pungent flavor and biting taste of mustard and Wassabi (Japanese horseradish). However, certain glucosinolates and their corresponding isothiocyanates have been found to exert potential benefits in inhibiting chemical and other types of carcinogenesis. For example, glucobrassicin and its myrosinase-catalyzed transformation products have been shown to inhibit carcinogenesis induced by polycyclic aromatic hydrocarbons and other initiators (Loft et al., 1992; Talalay and Zhang, 1996). An autolysis product of glucobrassicin, indole-3-carbinol, was found to induce monooxygenase (Loub et al., 1975) and glutathione-S-transferase (Sparnins et al., 1982), and inhibit development of forestomach, mammary, and pulmonary tumors (Wattenberg and Loub, 1978). Thus, potential benefits of certain glucosinolates may require the design of oilseeds where healthful glucosinolates are retained, or perhaps their content increased at the expense of their harmful counterparts.

OLIGOSACCHARIDES OF SUCROSE

These compounds, specifically raffinose, stachyose, and verbascose, are passed without digestion to the lower intestine where they are fermented (Naczk et al., 1997). This brings about generation of different gases which cause discomfort and cramps. However, these sugars are useful nutrients for the bifidus bacteria that reside in the lower intestine and are good for health. Therefore, potential benefits of these oligosaccharides is a subject of recent interest.

CYANOGENIC GLYCOSIDES

These are secondary metabolites which upon hydrolysis produce HCN. They are synthesized in plants from aromatic amino acids such as phenylalanine, tyrosine, and branched amino acids, i.e., leucine, isoleucine, and valine. Flaxseed contains several cyanogenic glycosides, i.e., linamarin, linustatin, and neolinustalin (Shahidi and Wanasundara, 1997). Other sources of cyanogenic glycosides are seeds of bitter almond and apricot, as well as those of peach and cherry. Cassava, lima bean, and bamboo may also contain cyanogens. Cyanogens in flaxseed meal have been shown to provide protection against selenium toxicity (Palmer, 1995).

CONCLUSIONS

Oilseeds, similar to other plant sources, provide an important reservoir of a myriad of phytochemicals. While some of the phytochemicals present may possess harmful effects when consumed in large amounts, the beneficial health effects of such components when used in appropriate quantities is well documented. To keep the balance, use of intact or crushed whole seeds, such as those of flax and sesame, together with other plant-based material, in food formulations and in baked products is recommended.

REFERENCES

Adlercreutz, H. 1997. Does fibre-rich food containing animal lignan precursors protect against both colon and breast cancer? An extention of the "fibre hypothesis," *Gastroenterology*, 86:761–764.

Adlercreutz, H. and Mazur, W. 1997. Phytoestrogens and Western diseases, *Ann. Med.* 29:95–120.

Amarowicz, R., Wanasundara, P.K.J.P.D., and Shahidi, F. 1994. Chromatographic separation of flaxseed phenolics, *Nahrung*, 4:520–526.

Bakke, J.E. and Klusterman, H.J. 1956. A new diglucoside from flaxseed, *Proc. N. Dakota Acad. Sci.*, 10:18–22.

Balasinska, B. 1998. Hypocholesterolemic effect of dietary evening primrose (*Oenothera paradoxa*) cake extract in rat, *Food Chem.*, 63:453–459.

Belitz, H.D. and Grosch, W. 1987. *Food Chemistry*, Springer Verlag, Berlin.

Blair, R. and Reichart, R.D. 1984. Carbohydrate and phenolic constitutents in a comprehensive range of rapeseed and canola fractions: nutritional significance for animals, *J. Sci. Food Agric.*, 35:29–35.

Blokland, A., Honig, W., Brouns, F., and Jolles, J. 1999. Cognition-enhancing properties of subcronic phosphatidylserine (PS) treatment in middle-aged rats: comparison of bovine cortex PS with egg PS and soybean PS, *Nutrition*, 15:778–783.

Bors, W., Hellar, W., Michel, C., and Saram, M. 1990. Flavonoids as antioxidants: determination of radical-scavenging efficiencies, *Meth. Enzymol.*, 186:343–351.

Boyd, L.C. 2001. Application of natural antioxidants in stabilizing polyunsaturated fatty acids in model systems and foods, in *Omega 3 Fatty Acids: Chemistry, Nutrition and Health Effects,* ACS Symp. Ser. 788, Shahidi, F., and Finley, J.W., eds., Am. Chem. Soc., Washington, D.C., pp. 258–279.

Cater, C.M., Gheyasuddin, S., and Mattil, K.F. 1972. The effect of chlorogenic, caffeic and quinic acids on the solubility and color of protein isolates, especially from sunflower seeds, *Cereal Chem.*, 49:508–514.

Chang, R.L., Huang, M.T., Wood, A.W., Wong, C.Q., Newmark, H.L., Yagi, H., Sayer, J.M., Jerina, D.M., and Conney, A.H. 1985. Effect of ellagic acid and hydroxylated flavonoids on the tumorigeniants of benzo (a) pyrene and (+) - 7 ß, 8α-dihydroxy - 9α, 10α-epoxy - 7,8,9,10-tetrahydrobenzo(a) pyrene on mouse skin and in newborn mice, *Carcinogenesis*, 6:1127–1133.

Choi, J.S., Yokozaiva, T., and Owa, H. 1991. Antihyperlipedemic effect of flavonoids from prunes deividiana, *J. Nat. Prod.*, 54:218–224.

Dabrowski, K.J. and Sosulski, F.W. 1984. Composition of free and hydrolyzable phenolic acids in defatted flours of ten oilseeds, *J. Agric. Food Chem.*, 32:128–130.

Dziezak, J.D. 1986. Antioxidants: the ultimate answer to oxidation, *Food Technol.*, 40(9):94–97, 100–102.

Foote, C.S., Chang, Y.C., and Denny, R.W. 1970. Chemistry of singlet oxygen. XI. *Cis-trans* isomerization of carotenoids by singlet oxygen and a possible quenching mechanism, *J. Am. Chem. Soc.,* 92:5218–5219.

Fukuda, Y., Osawa, T., Namiki, M., and Ozaki, T. 1985. Studies on antioxidative substances in sesame seeds, *Agric. Biol. Chem.,* 49:301–305.

Goh, S.H., Choo, Y.M., and Ong, S.H. 1985. Minor constituents of palm oil, *J. Am. Oil Chem. Soc.*, 62:237–240.

Halliwell, B. and Gutteridge, J.M.C. 1999. *Antioxidants in Nutrition, Health and Disease*, 3rd ed., Oxford University Press, Oxford, U.K.

Harland, B.F. and Obertas, D. 1977. A modified method for phytate analysis using an ion exchange procedure-application to textured vegetable proteins, *Cereal Chem.*, 54:827–836.

Hatano, T., Yasuhara, T., Matsuda, M., Yazaki, K., Yoshida, T., and Okuda, T. 1990. Oenothein B, a dimeric hydrolyzable tannin with macrocyclic structure and accompaning tannins from *Oenothera erythrosepala, J. Chem. Soc. Perkin Trans.*,1:2735–2743.

Hicks, K.B. and Moreau, R.A. 2001. Phytosterols and phytostenols: functional food cholesterol busters, *Food Technol.*, 55(1):63–67.

Hirota, A., Taki, S., Kawaii, S., Yano, M., and Abe, N. 2000. 1,1-Diphenyl-2-picrylhydrazyl radical-scavenging compounds from soybean miso and antiproliferative activity of isoflavones from soybean miso toward the cancer cell lines, *Biosci. Biotechnol. Biochem.*, 64:1038–1040.

Kamal-Eldin, A. and Appelqvist, L-A. 1996. The chemistry and antioxidant properties of tocopherols and tocotrienols, *Lipids*, 31:671–701.

Kozlowska, H., Zadernowski, R., and Sosulski, F.W. 1983a. Phenolic acids in oilseed flour, *Nahrung*, 27:449–453.

Kozlowska, H., Rotkiewicz, D.A., Zadornowski, R., and Sosulski, F.W. 1983b. Phenolic acids in rapeseed and mustard, *J. Am. Oil Chem. Soc.*, 60:1119–1123.

Krygier, K., Sosulski, F.W., and Hogge, C. 1982. Free, esterified, and insoluble-bound phenolic acids. III. Composition of phenolic acids in cereal and potato flours, *J. Agric. Food Chem.*, 30:337–340.

Kuenzig, W., Chan, J., Norkus, E., Holowaschenko, H., Newmark, H., Mergeus, W., and Conney, A.M. 1984. Caffeic and ferulic acids as blockers of nitrosamine formation, *Carcinogenesis*, 5:309–313.

Ladd, S.L., Sommer, S.A., La Berge, S., and Toscano, W. 1993. Effect of phosphatidylcholine on explicit memory, *Clin. Neuropharmacol.*, 16:540–549.

Lawhon, J.T., Carter, C.M., and Mattil, K.F. 1977. Evaluation of the food use potential of sixteen varieties of cottonseed, *J. Am. Oil Chem. Soc.*, 54:75–80.

Leung, J., Fenton, T.W., Mueller, M.M., and Claudinin, D.R. 1979. Condensed tannins of rapeseed meal, *J. Food Sci.*, 44:1313–1316.

Liener, I.E. 1997. Plant lectins: properties, nutritional significance and function, in *Antinutrients and Phytochemicals in Food,* ACS Symp. Ser. 662, Shahidi, F., ed., Am. Chem. Soc., Washington, D.C., pp. 31–43.

Liu, K. 2000. Expanding soybean food utilization, *Food Technol.*, 54(7): 46-48, 50, 52, 54, 56, 58.

Loft, S., Otte, J., Poulson, H.E., and Sorenson, H. 1992. Influence of intact and myrosinase-treated indolyl glucosinates on the metabolisms *in vivo* of metronidazole and antitryptine in rats, *Food Chem. Toxicol.*, 30:927–935.

Loub, W.D., Wattenberg, K.W., and Davis, D.W. 1975. Aryl hydrocarbon hydroxylase induction in rat tissues by naturally occurring indoles in cruciferous plants, *J. Natl. Cancer Int.*, 54:985–988.

Mazur, W. and Adlercreutz, H. 1998. Natural and anthropogenic environmental oestrogens: the scientific basis for risk assessment: naturally occurring oestrogens in food, *Pure Appl. Chem.*, 70:1759–1776.

Meck, W.H. and Williams, C.L. 1999. Choline supplementation during prenatal development reduces proactive interference in spatial memory, *Dev. Brain Res.*, 18:51–59.

Miyamoto, K.-I., Nomura, M., Sasakura, M., Matsui, E., Koshiura, R., Murayama, T., Furukawa, T., Hatano, T., Yoshida, T., and Okuda, T. 1993. Anti–tumor activity of oenothin B, a unique macrocyclic ellagiotannin, *Jpn. J. Cancer Res.*, 84:99–203.

Mori, H., Tanaka, T., Shima, H., Kuniyasu, T., and Takahashi, M. 1986. Inhibitory effect of chlorogenic acid on methylazoxymethanol acetate-induced carcinogenesis in large intestine and liver of hamsters, *Cancer Lett.*, 30:49–54.

Naczk, M. and Shahidi, F. 1989. The effort of methanol-ammonia-water/hexane treatment on the content of phenolic acids of canola, *Food Chem.*, 31:159–164.

Naczk, M., Nicols, T., Pink, D., and Sosulski, F.W. 1994. Condensed tannins in canola hulls, *J. Agric. Food Chem.*, 42:2196–2200.

Naczk, M., Amarowicz, R., and Shahidi, F. 1997. α-Galactosides of sucrose in foods: composition, flatulence-causing effects, and removal, in *Antinutrients and Phytochemicals in Food,* Shahidi, F., ed., ACS Symp. Ser. 662, Am. Chem. Soc., Washington, D.C., pp. 127–151.

Naczk, M., Amarowicz, R., Pink, D. and Shahidi, F. 2000. Insoluble condensed tannins of canola/rapeseed, *J. Agric. Food Chem.*, 48:1758–1762.

Namiki, M. 1995. The chemistry and physiological functions of sesame, *Food Rev. Int.*, 11:281–329.

Niki, E. 1991. Action of ascorbic acid as a scavenger of active and stable oxygen radicals, *Am. J. Clin. Nutr.*, 54:1119–1124.

Niki, E., Noguchi, N., Tsuchihashi, H., and Gotoh, N. 1995. Interaction among vitamin C, vitamin E, and beta-carotene, *J. Am. Clin. Nutr.*, 62:13225–13265.

Palmer, I.S. 1995. Effects of flaxseed on selenium toxicity, in *Flaxseed in Human Nutrition,* Cunnane, S.C. and Thompson, L.U., eds., AOCS Press, Champaign, IL, pp. 165–173.

Papas, A.M. 1999. Vitamin E: tocoopherols and tocotrienols, in *Antioxidant States, Diet, Nutrition and Health*, Papas, A.M., ed., CRC Press, Boca Raton, FL, pp. 189–210.

Piironen, V., Lindsay, D.G., Miettinsen, T.A., Toivo, J., and Lampi, A.-J. 2000. Plant sterols: biosynthesis, biological function and their importance to human nutrition, *J. Sci. Food Agric.*, 80:939–966.

Pratt, D.E. and Miller, E.E. 1984. A flavonoid antioxidant in Spanish peanuts, *J. Am. Oil Chem. Soc.*, 61:1064–1967.

Raj, A.S., Heddle, I.A., Newmark, H.L., and Katz, M. 1983. Caffeic acid as an inhibitor of DMBA-induced chromosomal breakage in mice assessed by bone-marrow micronucleus tests, *Mutat. Res.*, 124:247–253.

Rao, A.V. and Koratkar, R. 1997. Anticarcinogenic effects of saponins and phytosterols, in *Antinutrients and Phytochemicals in Food*, ACS Symp. Ser. 662, Shahidi, F., ed., Am. Chem. Soc., Washington, D.C., pp. 313–324.

Ratty, A.K. and Das, N.P. 1988. Effects of flavonoids on nonenzymatic lipid peroxidation: structure-activity relationship, *Biochem. Med. Metab. Biol.*, 39:67–79.

Rickard, S.E. and Thompson, L.U. 1997. Interactions and biological effects of phytic acid, in *Antinutrients and Phytochemicals in Food*, ACS Symp. Ser. 662, Shahidi, F., ed., Am. Chem. Soc., Washington, D.C., pp. 294–312.

Shahidi, F. 1997. *Natural Antioxidants: Chemistry, Health Effects and Applications*, AOCS Press, Champaign, IL.

Shahidi, F. 2000a. Antioxidant factors in plant foods and selected oilseeds, *BioFactors*, 13:179–185.

Shahidi, F. 2000b. Antioxidants in food and food antioxidants, *Nahrung*, 44:158-163.

Shahidi, F. and Naczk, M. 1995. *Food Phenolics: Sources, Chemistry, Effects and Applications,* Technomic Pub. Co., Lancaster, PA.

Shahidi, F. and Shukla, U.K.S. 1996. Nontriacylglycerol constituents of fats-oils, *Inform*, 7:1227–1232.

Shahidi, F. and Wanasundara, P.K.J.P.D. 1997. Cyanogenic glycosides of flaxseeds, in *Antinutrients and Phytochemicals in Food*, ACS Symp. Ser. 662, Shahidi, F., ed., Amer. Chem. Soc., Washington, D.C., pp. 171–185.

Shahidi, F., Wanasundara, U.N., and Amarowicz, R. 1994. Natural antioxidants from low pungency mustard flour, *Food Res. Int.*, 27:489–493.

Shahidi, F., Daun, J.K., and DeClercq, D.R. 1997. Glucosinolates in *Brassica* oilseeds: processing effects and extraction, in *Antinutrients and Phytochemicals in Food,* ACS Symp. Ser. 662, Shahidi, F., ed., Amer. Chem. Soc., Washington, D.C. pp. 152–170.

Shahidi, F., Wettasinghe, M., Amarowicz, R., and Khan, M.A. 2000. Antioxidants of evening primrose, in *Phytochemicals and Phytopharmaceuticals*, Shahidi, F. and Ho, C.-T., eds., AOCS Press, Champaign, IL., pp. 278–295.

Shukla, V.K.S., Wanasundara, P.K.J.P.D., and Shahidi, F. 1997. Natural antioxidants from oilseeds, in *Natural Antioxidants: Chemistry, Health Effects and Applications,* Shahidi, F., ed., AOCS Press, Champaign, IL., pp. 97–132.

Sims, R.J., Fioriti, J.A., and Kanuk, M.J. 1972. Sterol additives as polymerization inhibitors for frying oils, *J. Am. Oil Chem. Soc.*, 49:298–301.

Sparnins, V.L., Venogas, P.L., and Wattanberg, L.W. 1982. Glutathione-S–transferase activity: enhancement by dietary constituents, *J. Natl. Cancer Inst.*, 66:769–771.

Steinberg, D., Parthasarathy, S., Carew, T.E., Khoo, J.C., and Witztun, J.L. 1989. Beyond cholesterol. Modifications of low-density lipoprotein that increase its atherogenicity, *New Engl. J. Med.*, 320:915–924.

Talalay, P. and Zhang, Y. 1996. Chemoprotection against cancer by isothiocyanates and glucosinolates, *Biochem. Soc. Trans.*, 24:806–810.

Wanasundara, P.K.J.P.D. and Shahidi, F. 1994. Alcohol-ammonia-water/hexane extraction of flaxseed, *Food Chem.*, 49:39–44.

Wanasundara, U., Amarowicz, R., and Shahidi, F. 1994. Isolation and identification of an antioxidative component in canola, *J. Agric. Food Chem.*, 42:1285–1290.

Wattenberg, L.W. and Loub, W.D. 1978. Inhibition of polycyclic aromatic hydrocarbon–induced neoplasia by naturally occurring indoles, *Cancer Res.,* 38:1410–1413.

Wattenberg, L.W., Coccia, I.B., and Lam, L.K.T. 1980. Inhibitors effects of phenolic compound borzo (a) pyrene-indirect neoplasia, *Cancer Res.*, 40:2820–2823.

Westcott, N.D. and Muir, A.D. 1996. Process for extracting and purifying lignans and cinnamic acid derivatives from flaxseed, PCT Patent No. WO 9630468A2.

Westcott, N.D. and Muir, A.D. 2000. Overview of flaxseed lignans, *Inform,* 11:118-121.

Wettasinghe, M. and Shahidi, F. 1999a. Antioxidant and free radical-scavenging properties of ethanolic extracts of defatted borage (*Borago officinalis* L.) seeds, *Food Chem.,* 67:399–414.

Wettasinghe, M. and Shahidi, F. 1999b. Evening primrose meal: A source of national antioxidants and scavengers of hydrogen peroxide at oxygen-derived free radicals, *J. Agric. Food Chem.*, 47:1801–1812.

Wettasinghe, M. and Shahidi, F. 2000. Scavenging of reactive oxygen species and DPPH free radicals by extracts of borage and evening primrose meal, *Food Chem.*, 70:17–26.

Whitaker, J.R. 1997. Protease and α-amylase inhibitors in higher plants, in *Antinutrients and Phytochemicals in Food*, ACS Symp. Ser. 662, Shahidi, F., ed., Am. Chem. Soc., Washington, D.C., pp. 10–30.

White, P. and Xing, Y. 1997. Antioxidants from cereals and legumes, in *Natural Antioxidants: Chemistry, Health Effects and Applications*, Shahidi, F., ed., AOCS Press, Champaign, IL., pp. 25–63.

Wolf, W.J. 1983. Edible soybean protein products, in *Handbook of Processing and Utilization in Agriculture*, Wolf, I.A., ed., CRC Press, Boca Raton, FL, pp. 23–55.

Yan, P.S. and White, P.J. 1990. Linalyl acetate and other compounds with related structures as antioxidants in heated soybean oil, *J. Agric. Food Chem.*, 38:1904–1908.

CHAPTER 11

Lycopene and Human Health

BETTY JANE BURRI

CONTENTS

INTRODUCTION

Many people eat large amounts of lycopene, a powerful antioxidant found in tomatoes (Ford, 2000; Nebeling et al., 1997; Casso et al., 2000). Several interesting and recent experiments suggest that lycopene may protect against some forms of cancer and heart disease (Bramley, 2000; Gerster, 1997; Giovannucci, 1999; Arab and Steck, 2000). Cell culture studies show that lycopene inhibits the growth of many kinds of human cancer cells, sometimes more effectively than either alpha- or beta-carotene (Levy et al., 1995). Animal studies show that tumor formation can be inhibited by lycopene (Astorg et al., 1997; Nagasawa et al., 1995). Many human epidemiological and case-control studies suggest that tomatoes, or lycopene, can delay or prevent certain types of cancers (Giovannucci, 1999; Clinton, 1998). Currently, the most extensive evidence of protection appears to be against prostate cancer.

We know that these preliminary reports might paint an overly optimistic picture of the power of lycopene against disease. Often larger studies on the effect of nutrients on disease show lower efficacy than hoped. Prostate cancer is a common medical problem; therefore, even if lycopene simply postponed the progression of prostate cancer for a few years, this delay would be very useful. Any commonly consumed phytonutrient that can delay disease, pain, suffering, and medical costs by even a few years is well worth studying.

We have to be careful in devising nutrition studies for phytonutrients. A small but real benefit, or a benefit that occurs only for some people and not for others can be difficult and expensive to pinpoint. Hundreds of cell culture and epidemiological studies showed that people who ate more beta-carotene-rich fruits and vegetables had lower risks of premature death from cancer or heart disease than people who ate less (Ziegler, 1991). The overwhelmingly positive data from these cell-culture and epidemiological studies led to several clinical trials using beta-carotene supplements (Blot et al., 1995; Hennekens et al., 1996, Omenn et al., 1996; Alpha-Tocopherol Beta-Carotene Cancer Prevention Study Group, 1994). These trials were mostly disappointing. Many scientists now believe that either some other phytonutrient (such as lycopene) was protective, or that the maximum benefit from beta-carotene was provided from dosages that were much lower than given in the clinical trials (Burri et al., 1999; Lin et al., 1998). This paper reviews current research on lycopene and human health, and makes recommendations for future research directions in light of our experiences with beta-carotene.

LYCOPENE: WHAT IT IS, ITS FORMS, AND GOOD FOOD SOURCES

Lycopene is a bright red pigment that colors several fruits, vegetables, and flowers (Tables 11.1 and 11.2; Laval-Martin et al., 1975, Shi and Le Maguer, 2000). It has gained recent media attention because it may prevent or delay a variety of degenerative diseases, including cancer and heart disease (Gerster, 1997; Clinton, 1998; Krinsky, 1998; Rao and Agarwal, 2000). Chemically, lycopene is a carotenoid. There are over 700 members of the carotenoid family, but only a few of these are present in amounts that are easily measured in animal blood and tissue. The best known of these common carotenoids are beta-carotene, lycopene, and lutein. Lycopene is a long hydrocarbon chain with eleven conjugated double bonds, that lacks the characteristic ring structures found in most carotenoids (Figure 11.1) (Stahl and Sies, 1996; Sies and Stahl, 1998).

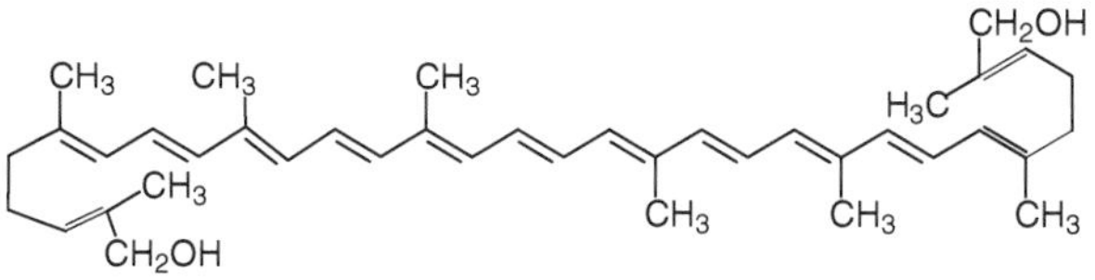

Figure 11.1 The chemical structure of lycopene.

ALL-TRANS LYCOPENE

There are many isomers of lycopene; an *all-trans* isomer and many varieties of *cis* isomer. *All trans*-Lycopene is the most common form of lycopene found in foods. In the common red tomato, about 90% of total lycopene is in this *all-trans* isomer (Beecher, 1998; Gartner et al., 1997; Hirota et al., 1982). *Trans*-Lycopene is also found in pink grapefruit, watermelon, and most other food sources of lycopene.

TABLE 11.1

Common Lycopene-Rich Fruits in the Diet

Tomato	*Lycopersicon esculentum*
Apricot	*Prunus armeniaca*
Grapefruit (pink)	*Citrus paradisi*
Guava (pink)	*Psidium guajava*
Watermelon	*Citrullus tanatis*

Minor food sources of lycopene in the American and European diet

Bitter melon	*Momordica charantia*
Carrot	*Daucus carota*
Cloudberry	*Rubus chamaemorus*
Cranberry	*Vaccinium vitis*
Date Palm	*Phoenix dactylifera*
Eggplant	*Solanum melongeria*
Grape	*Vitis vinifera*
Mango	*Mangifera indica*
Papaya	*Carica papaya*
Peach	*Prunus persica*
Sweet Pepper	*Capsicum annuum*
Persimmon	*Diospyros kaki*
Plum	*Prunus domestica*
Pumpkin	*Cucurbita pepo*
Rutabaga	*Brassica napus napobrassica*
Tea	*Camellia sinensis*
Turnip	*Brassica rapa rapa*

TABLE 11.2

Common Flowers Containing Lycopene

Calendula
Damask rose
Red bryony
Saffron
Rose
Yew

However, some foods contain *cis*-isomers. At least 90% of the lycopene found in tangerine tomatoes (*Lycoperison esculentum* var. Tangella) is the *tetra-cis* isomer (7Z, 9Z, 7'Z, 9'Z)-lycopene (Zechmeister,1941; Clough and Pattenden, 1979; Hirota et al., 1982). Many common forms in human blood and tissues are also *cis* isomers: 9-*cis*, 13-*cis*, and 15-*cis* lycopene.

Lycopene is found in a relatively small number of foods, most of which are pink or red fruits (Beecher, 1998; United States Department of Agriculture, 2000). Most of these fruits are native to the tropics (guava), or to the New World (http://www.tomato.org, 1/3/2002).

Lycopene was not a common phytonutrient in the diet of most people until modern times. However, millions of people eat lycopene regularly now. One of the richest sources of lycopene is also one of the world's largest crops, the tomato. California alone grows and processes over 11 million tons of tomatoes per year (http://www.tomato.org, 12/13/2000).

All forms of lycopene are brightly colored pigments, varying from orange to red. Aside from its presence in fruits and vegetables, lycopene tints the lobster shell, some fish scales and bird feathers, and some of the most colorful flowers found in nature.

POTENTIAL FUNCTIONS OF LYCOPENE IN THE HUMAN BODY

Like essential nutrients, lycopene is not made in the human body; however, unlike essential nutrients, lycopene does not seem to have a unique, irreplaceable function. Historically, millions of people survived without eating lycopene, since lycopene would have rarely been a part of pre-Columbian European and Asian diets. No known reduction of disease or increase in lifespan has been associated with the introduction of lycopene-rich foods into an area. No human study has shown a characteristic or identifiable physiological consequence from lycopene deprivation; however, this is not surprising, since no study has ever been designed to identify a characteristic deficit caused by lycopene deprivation. Stronger evidence is that no *in vitro* or *in vivo* experiment has shown any evidence for a unique or irreplaceable function. Lycopene is associated with antioxidant status (Sies and Stahl, 1998; Stahl et al., 1998), gap-junction formation (Bertram et al., 1991; Stahl et al., 2000), and inhibition of cholesterol synthesis (Aviram and Fuhrman, 1998; Fuhrman et al., 1997). We do not know whether other phytonutrients, such as beta-carotene, can replace lycopene, but it seems likely. Thus, it is very unlikely that lycopene is a classic, "essential nutrient."

Many phytonutrients that do not have unique, irreplaceable functions are important because they help preserve human health and well being. Hundreds of epidemiological studies have shown that people who eat greater than average amounts of phytonutrients from fruits and vegetables have lower risks of premature death from cancer and heart disease than people who eat fewer servings (Frei, 1995; Sengupta and Das, 1999; Ziegler, 1991). One of the largest and most consistent differences between people who eat greater or lesser amounts of fruit and vegetables is a difference in the amount of carotenoids they consume. And the carotenoid found in highest concentrations in people in the United States is lycopene (Ford, 2000).

Although most of the research on lycopene has centered on its antioxidant function, it also has a direct impact on cholesterol synthesis. Cholesterol homeostasis is maintained through feedback regulation of cholesterol biosynthesis and LDL-receptor activity. Cholesterol is synthesized by the mevalonate biosynthesis pathway. 3-hydoxy-3-methylglutaryl coenzyme A (HMGCoA) reductase controls the rate-limiting step in mevalonate biosynthetic pathway to cholesterol. Lycopene (and beta-carotene) are synthesized from mevalonate and share early synthesis steps with cholesterol. Therefore, these carotenoids could inhibit cholesterol synthesis, by competitively inhibiting the activity of HMGCoA reductase. Preliminary studies support the existence of this negative feedback loop: increased beta-carotene or lycopene concentrations suppress cholesterol synthesis and increase the removal of LDL from the plasma (Aviram and Furhman, 1998; Furhman et al., 1997).

Gap junctions function in cell-to-cell communication, by bridging cell walls and allowing the communication of small molecules between cells (Bertram et al., 1991; Krutovskikh et al., 1997; Stahl et al., 2000; Zhang et al., 1992). Connexin43, important in cell-to-cell communication, is initiated by a variety of carotenoids, including lycopene. The mechanism for this enhancement does not appear to be related to lycopene's antioxidant function, but instead is independent of it (Zhang et al., 1992). However, although gap-junction formation is necessary for normal cell growth and differentiation, lycopene does not appear to be a very potent initiator (Krutovskikh et al., 1997). Therefore, its role in gap-junction formation has not been established.

The first identified function of lycopene was immunological (Lingen et al., 1959). Research that is more recent has shown that lycopene can function in cholesterol metabolism and gap-junction formation, but none of these roles has been studied extensively. Most studies of lycopene and human health have focused on the antioxidant properties of lycopene (di Mascio et al., 1989; Lowe et al., 1999; Riso et al., 1999; Sies and Stahl, 1998; Stahl et al., 1998).

ANTIOXIDANTS IN HEALTH AND DISEASE

Lycopene is a powerful antioxidant (di Mascio et al., 1989; Sies and Stahl, 1998; Torbergsen and Collins, 2000). Although oxygen is essential to human life, it is also destructive. Rust and rot are both oxidative mechanisms. Many human tissues are also susceptible to oxidation, especially those containing polyunsaturated fats. Exposure to environmental pollutants such as smoking, smog, and irradiation increase oxidation. In fact, most scientists believe that oxidative damage is a major factor in most degenerative diseases of aging, such as heart disease, cancer, and macular degeneration. During life, our tissues would be susceptible to similar processes, except for the intervention of a variety of antioxidants. The antioxidant defense system is a network of vitamins (C, E), minerals (selenium, copper), phytonutrients (beta-carotene, lycopene, lutein), and biological products (bilirubin, coenzyme Q10) that protect tissues from oxidative damage (Jacob and Burri, 1996).

Lycopene is an especially powerful antioxidant because its multiplicity of conjugated double bonds makes it a good quencher of free radicals. Lycopene is also usually one of the most common carotenoids in the blood serum. Therefore, it can be an important part of the antioxidant defense system.

CELL CULTURE AND ANIMAL STUDIES THAT SHOW LYCOPENE MAY PREVENT CANCER

Recent experiments suggest that lycopene may function as an anticancer agent (Rauscher et al., 1998, Tinkler et al., 1994). *In vitro* studies showed that lycopene inhibited the growth of human leukemic (Amir et al., 1999), endometrial, lung, and mammary cancer cells (Karas et al., 2000; Lowe et al., 1999). In some cases, lycopene is more effective than either alpha- or beta-carotene (Levy et al., 1995, Tinkler et al., 1994). Animal studies have shown that liver (Astorg et al., 1997; Matsushima-Nishiwaki et al., 1995); brain, colonic (Narisawa et al., 1996; Narisawa et al., 1998), and mammary (Sharoni et al., 1997) tumorigenesis could be inhibited by lycopene (Bertram et al., 1991; Clinton, 1998). Experiments in rats and mice typically show that lycopene concentrates or pure lycopene can delay cancer progression. Of course, not all results reported have been positive. A lesser number have shown little or no benefit from lycopene supplements (e.g. Cohen et al., 1999).

HUMAN EPIDEMIOLOGICAL AND CASE-CONTROL STUDIES FOR LYCOPENE

Large epidemiological studies have shown that people who eat tomato products experience lower rates of cancer (Weisburger, 1998) and heart disease (Klipstein-Grobusch et al., 2000; Kohlmeier et al., 1997; Kristenson et al., 1997). Many epidemiological and case-control studies suggest that tomatoes, or lycopene, can delay or prevent certain types of cancers (Giovannuci et al., 1995; Jordan et al., 1997; de Stefani et al., 2000; Palan et al., 1996). Promising results have been found for cervical (Batieha et al., 1993) and breast (Dorgan et al., 1998; London et al., 1992) cancer.

The most studied interaction — and probably the most promising results — are against prostate cancer (Gann et al., 1999; Giovannucci et al., 1995; Hsing et al., 1990; Kristal and Cohen, 2000; Norrish et al., 2000; Rao et al., 1999). Estimates of lycopene consumption in these studies can be better predictors of prostate cancer risk than beta-carotene or total carotenoid consumption. Prostate cancer is a common and increasing medical problem (Carter and Coffey, 1990; Garnick and Fair, 1998), therefore, identifying a phytonutrient that may delay its onset would be useful.

There have been no clinical trials published on lycopene intervention, but several interventions have looked at preliminary biomarkers. For example, studies have reported that tomato consumption increased DNA protection from oxidation (Porrini and Riso 2000); and that tomato juice (Uprichard et al., 2000) or tomato oleoresin, a semipurified source of lycopene, decreased lipoprotein oxidation (Agarwal and Rao, 1998). Other studies have shown no effect from dietary supplements (Dugas et al., 1999) or tomato consumption (Pellegrini et al., 2000).

The fact that most lycopene in the diet comes from tomatoes and tomato products has had both beneficial and deleterious impact on research. It has been beneficial, because it is probably easier and more accurate to get good dietary intake estimates or food records when only a few foods are important. Furthermore, tomatoes are generally not hidden nutrients, as flour, salt, and sugar might be. Typically, when we eat tomato products we know that they contain tomatoes. Better and easier estimates of dietary intakes should lead to better nutrition research based on these estimates. Although lycopene is found in foods other than tomatoes, these foods are eaten rarely by most people. Substituting tomatoes for lycopene does not give a significantly biased result for people eating typical European or Western diets. Further, tomatoes are eaten by most ethnic groups, and cut across most cultures and lifestyles. For example, eating raw tomatoes in salads is associated with healthy lifestyles, while eating cooked tomato sauce on pizza (a better vehicle for lycopene) is associated with bad diets. So, unlike most phytonutrient-rich foods, tomatoes are probably not a surrogate for a healthy lifestyle, good income, increased education, or health consciousness.

Unfortunately many epidemiological studies have been conducted and reported as if 'tomato' and 'lycopene' were equivalent terms; however, a tomato is a complex entity, not a single phytonutrient. Tomatoes also provide substantial amounts of vitamin C and other carotenoids including beta-carotene, phytofluene, zeta-carotene, gamma-carotene, neurosporene, and phytoene. A recent study demonstrated that phytofluene is more bioavailable than expected based on its concentrations in tomatoes (Paetau et al., 1998). This suggests that phytofluene might play a role in protecting human health. Beta-carotene and vitamin C both have known biological functions. Furthermore, tomatoes are good sources of fiber and potassium. Thus, the functions hypothesized for lycopene from epidemiological studies may actually be associated with an unrelated phytonutrient.

BETA-CAROTENE: WHAT IT IS, GOOD FOOD SOURCES

Beta-carotene is a carotenoid with the many conjugated double bonds seen in lycopene, forming a connected double-ring structure. Beta-carotene plays a crucial role in human health, since it is the major source of vitamin A for most of the people throughout the world (Burri, 1997). It is a bright orange pigment found in a variety of vegetables and fruits, with a much wider distribution than lycopene. Although it is present in hundreds of dark-green vegetables, the most concentrated sources of beta-carotene are carrots, squash, pumpkin, and mangos (United States Department of Agriculture, 2000).

CELL CULTURE AND EPIDEMIOLOGICAL STUDIES OF BETA-CAROTENE, CANCER, AND HEART DISEASE

Over 250 epidemiological studies of cancer or heart disease risk have shown that high dietary intakes of beta-carotene rich fruits and vegetables were associated with lower risk of premature disease and death (Ziegler, 1991). Beta-carotene was linked to the prevention of cancer, heart disease, macular degeneration, and premature aging in these studies (Burri, 1997; Erdman et al., 1996). Cell culture, animal, human, and *in vitro* studies showed that beta-carotene was an effective antioxidant (Dixon et al., 1998; Dugas et al., 1999; Lin et al., 1998; Lowe et al., 1999). Furthermore, it was more powerful in gap-junction formation than lycopene (Bertram et al., 1991; Zhang et al., 1992), and inhibits cholesterol synthesis by the same mechanism as lycopene (Aviram and Fuhrman, 1998; Fuhrman et al., 1997).

BETA-CAROTENE AND THE CANCER TRIALS: CAUTIONS FOR LYCOPENE RESEARCH

The overwhelmingly positive data from these experimental and epidemiological studies led to a series of clinical trials using purified beta-carotene supplements (Alpha Tocopherol Beta Carotene Cancer Prevention Study Group, 1994, Blot et al., 1995; Hennekens et al., 1996; Omenn et al., 1996) in an effort to delay cancer incidence (Table 11.3); however, these clinical trials have been very disappointing. While some studies showed no difference, others had adverse effects.

TABLE 11.3
Beta Carotene Phase III Trials

Trial	Dosage (per day)	Results
PHS	25 mg BC	none
ATBC	30 mg BC	harm
CARET	30 mg BC + 25 mg retinoid	discontinued (potential harm)
Linxian	15 mg BC + Se + vitamin E	benefit

Note: PHS = Physician's Health Study (Hennekens, C.H., et al., *NEJM*, 1996; 334:1145. With permission.); ATBC = Alpha-Tocopherol Beta-Carotene Study (Alpha-Tocopherol Beta-Carotene Cancer Prevention Study Group, *NEJM*, 1994; 330:1029. With permission.); CARET = Carotene and Retinoid Study (Omenn, G.S., et al., *J. Natl. Cancer Inst.*, 1996; 88:1550. With permission.); Linxian = Linxian Study (Blot, W.J., et al., *Am. J. Clin. Nutr.*, 1995; 62:14245. With permission.); BC = beta-carotene, Se = selenium

Unfortunately, most clinical trials of beta-carotene were conducted by feeding high dosages of purified beta-carotene to well-fed individuals with easy access to beta-carotene-rich foods. For example, in the Physician's Health Study middle-aged and elderly male medical doctors ate a supplement containing approximately ten times the median dietary intake of beta-carotene, provided in a highly bioavailable form. No significant differences, positive or negative, between the treatment and control group occurred (Hennekens et al., 1996). Other clinical trials fed even larger amounts of beta-carotene, or beta-carotene plus retinol to Finnish smokers, or U.S. smokers and asbestos workers (Alpha Tocopherol Beta Carotene Cancer Prevention Study Group, 1994; Omenn et al., 1996). These trials indicated that high doses of beta-carotene were harmful. Smokers and asbestos workers who ate the beta-carotene supplements had higher rates of cancer than the control group. However, a clinical trial of a mixed antioxidant supplement containing lower dosages of beta-carotene (plus selenium and vitamin E) given to marginally malnourished adults in Linxian China showed positive results (Blot et al., 1995). In this poorly fed group of adults, moderate antioxidant supplementation decreased cancer risk and death.

The dosage of beta-carotene selected, the other nutrients provided, and the populations studied may all have made a difference between success and failure in these trials. In fact, clinical trials for many nutrients are problematic, because they may give significant results for one population and not another. Nutrient interventions generally only give good results if the amount of the nutrient that is beneficial is higher than typical dietary intakes. Studies that deplete people of nutrients can be difficult and expensive, but they identify physiologically important functions with more certainty.

CAROTENOID DEPLETION STUDIES

Carotenoid depletion trials, where well-fed people are fed foods low in lycopene and other carotenoids, should give a clearer picture of whether, and when, carotenoids such as lycopene might have important physiological functions. Unfortunately, there have only been a few studies of carotenoid depletion on oxidative damage, and none on gap-junction formation or cholesterol metabolism. All of these studies have shown deleterious increases in oxidative damage during depletion (Burri, 1997; Burri et al., 1999; Dixon et al., 1998; Lin et al., 1998).

We used the information from a well-controlled double-blind placebo-controlled carotenoid-depletion study to estimate the effective range of beta-carotene in the antioxidant defense system. The study population was healthy adult women who lived on our metabolic unit for 120 days. They ate foods low in carotenoids, but that provided adequate nutrients and energy. We measured serum retinol and carotenoid concentrations, and conducted several tests to assess oxidative damage (malondialdehyde-thiobarbituric acid concentrations and reactive carbonyls). We analyzed results by fitting a random regression coefficient model to the data. The saturation point was extrapolated similarly to a low-dose extrapolation in dose-response analysis. The maximal protection of low-density lipoproteins from oxidative damage occurred at serum concentrations of 2.3 ± 1.6 micromoles/L. These serum concentrations are provided by intakes of 10 to 18 micromoles/day of carotenoid (Lin et al., 1998). Thus, maximal protection from oxidative damage occurs at relatively low carotenoid concentrations that are easily attained by eating the recommended number of servings of fruits and vegetables per day. Higher dosages of carotenoids (15 mg/day, about half the dosage used in the Phase III trials) normalized oxidative damage in depleted subjects but did not improve it over baseline values.

Our results suggest that maximal protection against oxidative damage is provided by relatively modest intakes of carotenoids. It follows that carotenoid trials that feed small amounts of beta-carotene to people who eat diets low in fruits and vegetables (similar to the Linxian trial; Blot et al., 1995) may be more successful than the clinical trials that fed large amounts of beta-carotene to well-fed people.

FUTURE DIRECTIONS FOR LYCOPENE STUDIES

These results have implications for lycopene research. The most obvious implication relates to any clinical trial planned to test the effect of lycopene supplements on cancer or heart disease. Normally, Phase III clinical trials are considered the gold standard of human research experiments; however, clinical trials were designed to test new drugs. Ideally, the control group in a clinical trial has never taken the drug, has no access to it, and will not be able to take it during the trial. Meanwhile, the intervention group takes the drug in amounts that have already been estimated to be safe and effective in smaller studies. This is not possible for most phytonutrients, including lycopene. A plentiful source of lycopene—tomatoes—is inexpensive, available, and commonly eaten in large quantities already. The control groups available for most clinical studies of lycopene, then, has eaten and will continue to eat variable and sometimes very large amounts of lycopene on an irregular but frequent basis. To get good comparisons between the control and treatment groups, the treatment group would have to eat enough lycopene to raise its levels high above background lycopene intakes. Further, the amount of lycopene fed to the treatment group might have to be based on an amount of lycopene large enough to separate it from the control group, and not on scientific estimates of the safe and effective dose. Similar to beta-carotene, experiments already suggest that lycopene may be an effective antioxidant only at relatively low concentrations (Lowe et al., 1999).

Thus, clinical trials are likely to show positive results for lycopene only if the amount of lycopene that is beneficial is significantly and substantially more than the typical baseline intakes of lycopene. In many communities in the U.S. and throughout the world, these intakes would have to be very high; several large servings of tomatoes a week. There is no convincing evidence that these high dosages are necessary to attain any health benefits lycopene could provide. Unfortunately, the higher the intake of lycopene in the treatment group, the more likely the dosage will exceed its safe upper limit of intake. It might be given at such high levels in clinical trials that deleterious side effects of the high dosage would predictably outweigh the benefits.

A broader implication is that forcing useful but nonessential nutrients like lycopene to meet the definitions and tests set up for essential nutrients can give bad results, and hamper research and education. Experiments that worked well for essential nutrients may not be suitable for nutrients that are useful, but replaceable. If a nutrient is useful but not essential, if it can be replaced, then it will be replaced in some people. Thus, some people will benefit from the phytonutrient, while others will not.

We know that lycopene appears to be useful, but not essential. However, we do not know yet what nutrients, or combination of nutrients, might replace it. We do not know how much lycopene is useful, or under what circumstances these amounts may change. We have not identified the populations that might benefit from lycopene. Finally, we do not know how completely lycopene may be replaced, how simply and easily. To use an analogy: a person who loses their right hand can learn to use their left hand for all critical functions, so that it would be difficult to prove scientifically that they needed their right hand. Still, two hands make life's functions much simpler and easier. Similarly, lycopene could be replaced by other phytonutrients to the extent that no essential functions could be reliably identified through scientific experiments, yet still not be entirely and reasonably replaced.

Suppose all the functions normally done by lycopene could be replaced by an equivalent amount of beta-carotene. That substitution is simple and reasonable. We could make dietary recommendations for carotenoid-rich foods, instead of making separate recommendations for beta-carotene-rich foods and lycopene-rich foods. On the other hand, suppose all the functions could be replaced, but only with ten times as much beta-carotene, plus increased vitamin E and vitamin C. The replacement is neither simple nor especially reasonable. It would usually be easier to recommend lycopene-rich foods instead of the foods that could replace it.

Lycopene, lutein, wine solids, garlic and onions, soy, and many other phytonutrients may be beneficial to some populations in physiological concentrations. Our experience with beta-carotene suggests we should treat this phytonutrient with cautious optimism; however, we need to improve and target our phytonutrient intervention studies to get better results. More basic research on what lycopene does, how much lycopene appears to be beneficial, who might benefit, and what other phytochemicals can replace it are critical. Ideally, this should be determined before many more lycopene intervention trials are planned. Hopefully, future trials will also explore more mechanisms than oxidative damage, and will investigate the role of lycopene in cholesterol synthesis and gap-junction formation. Meanwhile, setting up clinical interventions with relatively low supplements of lycopene (similar to that provided by a slice of pizza or serving of tomato sauce) targeted to people eating low amounts of carotenoids are unlikely to do harm and may be beneficial.

REFERENCES

Agarwal, S. and Rao, A.V. Tomato lycopene and low-density lipoprotein oxidation: a human intervention study, *Lipids,* 1998; 33:981–984.

Alpha Tocopherol Beta Carotene Cancer Prevention Study Group. The effect of vitamin E and beta-carotene on the incidence of lung cancer and other cancers in male smokers, *N. Eng. J. Med.*, 1994; 330:1029–1035.

Amir, H., Karas, M., Giat, J., Danilenko, M., Levy, R., Yermiahu, T., Levy, J., and Sharoni, Y. Lycopene and 1,25-dihydroxyvitamin D3 cooperate in the inhibition of cell cycle progression and induction of differentiation in HL-60 leukemic cells, *Nutr. Cancer,* 1999; 33:105–112.

Arab, L. and Steck, S. Lycopene and cardiovascular disease, *Am. J. Clin. Nutr.,* 2000; 71:1691S–1695S.

Astorg, D., Gradelet, S., Berges, R., and Suschetet, M. Dietary lycopene decreases the initiation of liver preneoplastic foci by diethylnitrosamine in the rat, *Nutr. Cancer,* 1997; 29:60–68.

Aviram, M. and Fuhrman, B. LDL oxidation by arterial wall macrophages depends on the oxidative status in the lipoprotein and in the cells: role of prooxidants vs. antioxidants, *Mol. Cell Biochem.*, 1998; 188:149–159.

Batieha, A.M., Armenian, H.K., Norkus, E.P., Morris, J.S., Spate, V.E., and Comstock, G.W. Serum micronutrients and the subsequent risk of cervical cancer in a population–based nested case-control study, *Cancer Epidemiol. Biomark. Prev.,* 1993; 2:335–339.

Beecher, G.R. Nutrient content of tomatoes and tomato products, *Proc. Soc. Exp. Biol. Med.* 1998; 218:98–100.

Bertram, J.S., Pung, A., Churley, M., Kappock, T.J., Wilkins, L.R., and Cooney, R.V. Diverse carotenoids protect against chemically induced neoplastic transformation, *Carcinogenesis*, 1991; 12:671–678.

Blot, W.J., Li, J., Taylor, P.R., Guo, W., Dawsey, S.M., and Li, B. The Lixian trials: mortality rates by vitamin-mineral intervention group, *Am. J. Clin. Nutr.*, 1995; 62:1424S–1426S.

Bramley, P.M. Is lycopene beneficial to human health? *Phytochemistry,* 2000; 54:233–236.

Burri, B.J. Beta-carotene and human health: a review of current research, *Nutr. Rev.,* 1997; 17:547–580.

Burri, B.J., Clifford, A.J., and Dixon, Z.R. Beta-carotene depletion and oxidative damage in women, *R. Soc. Chem.* (Scand.), Natural antioxidants and anticarcinogens in nutrition, health and disease, Kumulainen, J.T. and Salonen, J.T., eds., pp. 231–234. 1999.

Carter, H.B. and Coffey, D.S. The prostate: an increasing medical problem, *Prostate*, 1990; 16:39–48.

Casso, D., White, E., Patterson, R.E., Agurs-Collins, T., Kooperberg, C., and Haines, P.S. Correlates of serum lycopene in older women, *Nutr. Cancer,* 2000; 36:163–169.

Clinton, S.K. Lycopene: chemistry, biology, and implications for human health and disease, *Nutr. Rev.*, 1998; 56:35–51.

Clough, J.M. and Pattenden, G. Naturally occurring *poly-cis* carotenoids. Stereochemistry of *poly-cis* lycopene and its congeners in Tangerine tomato fruits, *JCS Chem. Comm.*, 1979; 616–619.

Cohen, L.A., Zhao, Z., Pittman, B., and Khachik, F. Effect of dietary lycopene on N-methyl-nitrosourea-induced mammary tumorigenesis, *Nutr. Cancer,* 1999; 34:153–159.

De Stefani, E., Oreggia, F., Boffetta, P., Deneo-Pellegrini, H., Ronco, A., and Mendilaharsu, M. Tomatoes, tomato-rich foods, lycopene and cancer of the upper respiratory tract: a case control in Uruguay, *Oral Oncol.*, 2000; 36:47–53.

Di Mascio, P., Kaiser, S., and Sies, H. Lycopene as the most efficient biological carotenoid singlet oxygen quencher, *Arch. Biochem. Biophys.*, 1989; 274:532–538.

Dixon, Z.R., Feng-Shiun, S., Warden, B.A., Burri, B.J., and Neidlinger, T.R. The effect of a low carotenoid diet on malondialdehyde-thiobarbituric acid (MDA-TBA) concentrations in women: a placebo-controlled double-blind study, *J. Am. Coll. Nutr.*, 1998; 17:54–58.

Dorgan, J.F., Sowell, A., Swanson, C.A., Potischman, N., Miller, R., Schussler, N., and Stephensen, H.E., Jr. Relationships of serum carotenoids, retinol, alpha-tocopheral, and selenium with breast cancer risk: Results from a prospective study in Columbia, MO, *Cancer Causes Control*, 1998; 9:89–97.

Dugas, T.R., Morel, D.W., and Harrison, E.H. Dietary supplementation with beta-carotene, but not with lycopene, inhibits endothelial cell-mediated oxidation of low density lipoprotein, *Free Rad. Biol. Med.*, 1999; 26:1238–1244.

Erdman, J.W., Jr., Russell, R.M., Rock, C.L., Barua, A.B., Bowen, P.E., Burri, B.J., Curran-Celentano, J., Furr, H., Mayne, S.T., and Stacewicz-Sapuntzakis, M. Beta-carotene and the carotenoids: beyond the intervention trials, *Nutr. Rev.*, 1996; 54:185–188.

Ford, E.S. Variations in serum carotenoid concentrations among United States adults by ethnicity and sex, *Ethn. Dis.*, 2000;10:208–217.

Frei, B. Cardiovascular disease and nutrient antioxidants: role of low-density lipoprotein oxidation, *Crit. Rev. Food Sci. Nutr.*, 1995; 35:83–98.

Furhman, B., Elis, A., and Aviram, M. Hypocholesterolemic effect of lycopene and beta-carotene is related to suppression of cholesterol synthesis and augmentation of LDL receptor activity in macrophages, *Biochem. Biophys. Res. Comm.*, 1997; 233:658–662.

Gann, P.H., Ma, J., Giovannucci, E., Willett, W., Sacks, F.M., Hennekens, C.H., and Stampfer, M.J. Lower prostate cancer risk in men with elevated plasma lycopene: results of a prospective analysis, *Cancer Res.*, 1999; 59:1225–1230.

Garnick, M.B. and Fair, W.R. Combating prostate cancer, *Sci. Am.*, 1998; 279:74–83.

Gartner, C., Stahl, W., and Sies, H. Lycopene is more available from tomato paste than from fresh tomatoes, *Am. J. Clin. Nutr.*, 1997; 66:116–122.

Gerster, H. The potential role of lycopene for human health, *J. Am. Col. Nutr.*, 1997;16:109–126.

Giovannucci, E. Tomatoes, tomato-based products, lycopene, and cancer: review of epidemiological literature, *J. Natl. Cancer Inst.*, 1999; 91:317–331.

Giovannucci, E., Ascherio, A., Rimm, E.B., Stampfer, M.J., Colditz, G.A., and Willett, W.C. Intake of carotenoids and retinol in relation to risk of prostate cancer, *J. Nat. Cancer Inst.* 1995; 87:1767–1776.

Hennekens, C.H., Buring, J.E., Manson, J.E., Stampfer, M., Rosner, B., Cook, N.R., Belanger, C., La Motte, F., Gaziano, J.M., Ridker, P.M., Willett, W., and Peto, R. Lack of effect of long-term supplementation with beta-carotene on the incidence of malignant neoplasms and cardiovascular disease, *NEJM*, 1996; 334:1145–1149.

Hirota, S., Sato, H., and Tsuyuki, H. The carotenoid constitution of various tomato strains. IV. Separation and determination of carotenoids in fruit flesh of typical tomato strains, *Nippon Shokuhin Kogyo Gakkaishi*, 1982; 29:477–483.

Hsing, A.W., Comstock, G.W., Abbey, H., and Polk, B.F. Serological precursors of cancer: retinol, carotenoids, and tocopherol, and the risk of prostate cancer, *J. Natl. Cancer Inst.*, 1990; 82:941–946.

Jacob, R.A. and Burri, B.J. Oxidative damage and defense, *Am. J. Clin. Nutr.*, 1996; 63:985S–990S.

Jordan, P., Brubacher, D., Tsugane, S., Tsubono, Y., Gey, K.F., and Moser, U. Modeling of mortality data from a multi-centre study in Japan by means of Poission regression with error in variables, *Int. J. Epidemiol.*, 1997; 26: 501–507.

Karas, M., Amir, H., Fishman, D., Danilenko, M., Segal, S., Nahum, A., Koifmann, A., Giat, Y., Levy, J., and Sharoni, Y. Lycopene interferes with cell cycle progression and insulin-like growth factor I signalling in mammary cancer cells, *Nutr. Cancer,* 2000; 36: 101–111.

Klipstein-Grobusch, K., Launer, L.J., Geleijnse, J.M., Boeing, H., Hofman, A., and Witteman, J.C. Serum carotenoids and atherosclerosis: the Rotterdam Study, *Athersclerosis*, 2000; 148: 49–56.

Kohlmeier, L., Kark, J.D., Gomez-Gracia, E., Martin, B.C., Steck, S.E., Kardinaal, A.F., Ringstad, J., Thamm, M., Masaev, V., Riemersma, R., Martin-Moreno, J.M., Huttunen, J.K., and Kok, F.J. Lycopene and myocardial infarction risk in the EURAMIC Study, *Am. J. Epidemiol.*, 1997; 146:618–626.

Krinsky, N.I. Overview of lycopene, carotenoids, and disease prevention, *Proc. Soc. Exp. Biol. Med.*, 1998; 218:95–97.

Kristal, A.R. and Cohen, J.H. Invited commentary: tomatoes, lycopene, and prostate cancer. How strong is the evidence, *Am. J. Epidemiol.*, 2000; 151:124–127.

Kristenson, M., Zieden, B., Kucinskiene, Z., Elinder, L.S., Bergdahl, B., Elwing, B., Abaravicius, A., Razinkoviene, L., Calkauskas, H., and Olsson, A.G. Antioxidant state and mortality from coronary heart disease in Lithuanian and Swedish men: concomitant cross sectional study of men aged 50, *BMJ,* 1997; 314:629–633.

Krutovskikh, V., Asamoto, M., Taksuka, N., Murakoshi, M., Nishino, H., and Tsuda, H. Differential dose-dependent effects of alpha-, beta-carotenes and lycopene on gap-junctional intercellular communication in rat liver *in vivo, Jpn. J. Cancer Res.*, 1997; 88:1121–1124.

Laval-Martin, D., Quennemet, J., and Moneger, R. Pigment evolution of *Lycopersican esculentum* fruits during growth and ripening, *Phytochemistry*, 1975; 14:2357–2362.

Levy, J., Bosnin, E., Feldman, B., Giat, Y., Miinster, A., Danilenko, M., and Sharoni, Y. Lycopene is a more potent inhibitor of human cancer cell proliferation than either alpha-carotene or beta-carotene, *Nutr. Cancer,* 1995; 24:257–266.

Lin, Y., Burri, B.J., Neidlinger, T.R., Muller, H.-G., Dueker, S.R., and Clifford, A.J. Estimating the concentration of beta-carotene required for maximal protection of low-density lipoproteins in women, *Am. J. Clin. Nutr.*, 1998; 67:837–845.

Lingen, C., Ernster, L., and Lindberg, O. The promoting effect of lycopene on the nonspecific resistance of animals, *Exp. Cell Res.,* 1959; 16:384–393.

London, S.J., Stein, E.A., Henderson, I.C., Stampfer, M.J., Wood, W.C., Remine, S., Dmochowski, J.R., Robert, N.J., and Willett, W.C. Carotenoids, retinol, and vitamin E and risk of proliferative benign breast disease and breast cancer, *Cancer Causes Control,* 1992; 3:503–512.

Lowe, G.M., Booth, L.A., Young, A.J., and Bilton, R.F. Lycopene and beta-carotene protect against oxidative damage in HT29 cells at low concentrations but rapidly lose this capacity at high doses, *Free Radic. Res.*, 1999; 30:141–151.

Matsushima-Nishiwaki, R., Shidoji, Y., Nishiwaki, S., Yamada, T., Moriwaki, H., and Muto, Y. Suppression by carotenoids of microcystin-induced morphological changes in mouse hepatocytes, *Lipids*, 1995; 30:1029–1034.

Nagasawa, H., Mitamura, T., Sakamoto, S., and Yamamoto, K. Effects of lycopene on spontaneous mammary tumor development in SHN virgin mice, *Anticancer Res.*, 1995; 15:1173–1178.

Narisawa, T., Fukaura, Y., Hasebe, M., Ito, M., Aizawa, R., Murakoshi, M., Uemura, S., Khachik, F., and Nishino, H. Inhibitory effects of natural carotenoids, alpha-carotene, beta-carotene, lycopene, and lutein, on colonic aberrant crypt foci formation in rats, *Cancer Lett.*, 1996; 107:137–142.

Narisawa, T., Fukaura, Y., Hasebe, M., Nomura, S., Oshima, S., Sakamoto, H., Ishiguro, Y., Takayasu, J., and Nishino, H. Prevention of N-methynitrosourea-induced colon carcinogenesis in rats by lycopene and tomato juice rich in lycopene, *Jpn. J. Cancer Res.,* 1998; 89:1003–1008.

Nebeling, L.C., Forman, M.R., Graubard, B.I., and Snyder, R.A. Changes in carotenoid intake in the United States: the 1987 and 1992 National Health Interview Surveys, *J. Am. Diet Assoc.*, 1997; 97:991–996.

Norrish, A.E., Jackson, R.T., Sharpe, S.J., and Skeaff, C.M. Prostate cancer and dietary carotenoids, *Am. J. Epidemiol.,* 2000; 151:119–123.

Omenn, G.S., Goodman, G.E., Thornquist, M.D., Balmes, J., Cullen, M.R., Glass, A., Keogh, J.P., Meyskens, F.L., Jr., Valanis, B., Williams, J.H., Jr., Barnhardt, S., Cherniack, M.G., Brodkin, C.A., and Hammar, S. Risk factors for lung cancer and for intervention effects in CARET, the Beta-Carotene and Retinol Efficacy Trial, *J. Natl. Cancer Inst.,* 1996; 88:1550–1559.

Paetau, I., Khachik, F., Brown, E.D., Beecher, G.R., Kramer, T.R., Chittams, J., and Clevidence, B.A. Chronic injestion of lycopene-rich tomato juice or lycopene supplements significantly increase plasma concentrations of lycopene and related tomato carotenoids in humans, *Am. J. Clin. Nutr.,* 1998; 68:1187–1195.

Palan, P.R., Mikhail, M.S., Goldberg, G.L., Basu, J., Runowicz, C.D., and Romney, S.L. Plasma levels of beta-carotene, lycopene, canthaxanthin, retinol, and alpha- and tau-tocopherol in cervical intra epithelial neoplasm and cancer, *Clin. Cancer Res.*, 1996; 2:181–185.

Pellegrini, N., Riso, P., and Porrini, M. Tomato consumption does not affect the total antioxidant capacity of plasma, *Nutr.,* 2000; 16:268–271.

Porrini, M. and Riso, P. Lymphocyte lycopene concentration and DNA protection from oxidative damage is increased in women after a short period of tomato consumption, *J. Nutr.,* 2000; 130:189–192.

Rao, A.V. and Agarwal, S. Role of antioxidant lycopene in cancer and heart disease, *J. Am. Col. Nutr.*, 2000; 19:563–569.

Rao, A.V., Fleshner, N., and Agarwal, S. Serum and tissue lycopene and biomarkers of oxidation in prostate cancer patients: a case control study, *Nutr. Cancer,* 1999; 33:159–164.

Rauscher, R., Edenharder, R., and Platt, K.L. *In vitro* antimutagenic and *in vivo* anticlastogenic effects of carotenoids and solvent extracts from fruits and vegetables rich in carotenoids, *Mutat. Res.*, 1998; 413:129–142.

Riso, P., Pinder, A., Santangelo, A., and Porrini, M. Does tomato consumption effectively increase the resistance of DNA to oxidative damage? *Am. J. Clin. Nutr.*, 1999; 69:712–718.

Sengupta, A. and Das, S. The anti-carcinogenic role of lycopene, abundantly present in tomato, *Eur. J. Cancer Prev.*, 1999; 8:325–330.

Sharoni, Y., Giron, E., Rise, M., and Levy, J. Effects of lycopene-enriched tomato oleoresin on 7,12-dimethyl-benz [a] anthracene-induced rat mammary tumors, *Cancer Detect. Prev.,* 1997; 21:118–123.

Shi, J. and Le Maguer, M. Lycopene in tomatoes: chemical and physical properties affected by food processing, *Crit. Rev. Food Sci. Nutr.*, 2000; 40:1–42.

Sies, H. and Stahl, W. Lycopene: antioxidant and biological effects and its bioavailability in the human, *Proc. Soc. Exp. Biol. Med.,* 1998; 218:121–124.

Stahl, W., Junghans, A., de Boer, B., Driomina, E.S., Briviba, K., and Sies, H. Carotenoid mixtures protect multilamellar liposomes against oxidative damage: synergistic effects of lycopene and lutein, *FEBS Lett.*, 1998; 427:305–308.

Stahl, W. and Sies, H. Lycopene: a biologically important carotenoid for humans? *Arch. Biochem. Biophys.,* 1996; 336:1–9.

Stahl, W., von Laar, J., Martin, H.D., Emmerich, T., and Sies, H. Stimulation of gap junctional communication: comparison of acyclo-retinoic acid and lycopene, *Arch. Biochem. Biophys.,* 2000; 373:271–274.

Tinkler, J.H., Bohm, F., Schalch, W., and Truscott, T.G. Dietary carotenoids protect human cells from damage, *J. Photochem. Photobiol. B.*, 1994; 26:283–285.

Torbergsen, A.C. and Collins, A.R. Recovery of human lymphocytes from oxidative DNA damage; enhancement of DNA repair by carotenoids is probably simply an antioxidant effect, *Eur. J. Nutr.,* 2000; 39:80–85.

Uprichard, J.E., Sutherland, W.H., and Mann, J.I. Effect of supplementation with tomato juice, vitamin E, and vitamin C on LDL oxidation and products of inflammatory activity in type 2 diabetes, *Diabetes Care,* 2000; 23:733–738.

United States Department of Agriculture. Carotenoid database for U.S. foods 1998. http://www.nal.usda.gov/fnic/foodcomp/Data/car98/car98html. December 22, 2000.

Weisburger, J.H. Evaluation of the evidence on the role of tomato products in disease prevention, *Proc. Soc. Exp. Biol. Med.,* 1998; 218:140–143.

Zechmeister, L., Le Rosen, A.L., West, F.W., and Pauling, L. Prolycopene, a naturally occurring sterioisomer of lycopene, *Proc. Natl. Acad. Sci. U.S.A.*, 1941; 2:468–471.

Zhang, L.X., Cooney, R.V., and Bertram, J.S. Carotenoids up-regulate connexin43 gene expression independent of their provitamin A or antioxidant properties, *Cancer Res.,* 1992; 52:5707–5712.

Ziegler, R.G. Vegetables, fruits, and carotenoids and the risk of cancer, *Am. J. Clin. Nutr.,* 1991; 53:S251–S259.

CHAPTER 12

Astaxanthin, β-Cryptoxanthin, Lutein, and Zeaxanthin

JOHN T. LANDRUM,
RICHARD A. BONE,
and CHRISTIAN HERRERO

CONTENTS

INTRODUCTION

Carotenoids are predominantly produced by photosynthetic plants, algae, bacteria, and some fungi (Britton et al., 1995; Weedon, 1971). The functions of carotenoids in photosynthesis are: 1) to intercept and quench the excited triplet-state chlorophyll molecule preventing the generation of singlet oxygen and 2) to serve as accessory pigments absorbing light in the blue region of the visible light spectrum where chlorophyll is an inefficient absorber (Goodwin, 1980; Krinsky, 1971). In higher plants, carotenoids function in the zeaxanthin cycle, which is responsible for regulating the extent that blue light is converted into chemical energy by the photosynthetic process (Demmig-Adams et al., 1996). A host of other essential or subsidiary functions are known or postulated for carotenoids in plants, including the important functions of leaf, flower, and fruit coloration (Goodwin, 1980). The process of evolution has refined photosynthesis for a period of more than one billion years during which a wide diversity of naturally occurring carotenoids have developed specialized roles in the many plant and algal species. Nearly 1000 structurally distinct carotenoids that are natural products have been isolated from living systems (Straub, 1987). The total global biosynthesis of carotenoids is estimated to be in excess of 100 million tons per year (Britton et al., 1995). Higher animals are unable to synthesize carotenoids but, nevertheless, have developed a dependence on these compounds for a range of functions.

Primarily because of the limited number of carotenoids that are present in the dominant food plants, higher animals and humans consume only a small fraction of the known natural carotenoid structural types (Khachik et al., 1991). In human serum there are as many as 50 carotenoids, but only half a dozen or so are normally present in percentages exceeding 5–10% (Khachik et al., 1995; 1992). These include α- and β-carotene, lycopene, β-cryptoxanthin, lutein, and zeaxanthin. Figure 12.1 shows the typical composition of the carotenoids extracted from human serum in the U.S. These carotenoids are all fairly abundant in varying quantities in the green and yellow vegetables, and red and orange fruits (Klaüi and Bauernfeind, 1981). The hydrocarbon carotenoids, α- and β-carotene, and lycopene have received intense scrutiny because of their abundance and potential or proven functions in animal physiological processes, including provitamin-A status. It is the xanthophylls, oxycarotenoids, that are the focus of our attention.

β-Cryptoxanthin, lutein, and zeaxanthin are the major, although not exclusive, oxycarotenoid components of human serum. With the exception of β-cryptoxanthin, these are not provitamin A carotenoids because of the functional oxygen groups present in their structures. There is a keen interest in defining what their functional significance may be. It is important to understand to what extent these compounds individually or collectively contribute to optimal human health, and whether they do so by uniquely serving one or more vital physiological functions within one or more tissues. We include in our discussion the naturally abundant oxycarotenoid astaxanthin, and while it is a minor or perhaps more accurately an occasional component in the human diet, it has been identified as an especially effective antioxidant (Martin et al., 1999).

A question that naturally arises as part of this discussion is: What criteria establish when a dietary component, an oxycarotenoid in this instance, serves an essential functional role within a biological system? We might also ask: What criteria establish when a nonessential dietary component confers benefits that are significant to optimal health? Clearly, there are simplistic answers to each question, but it is

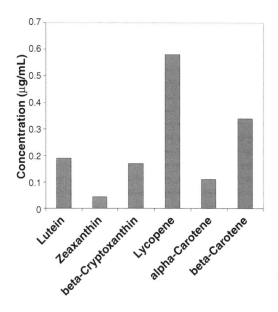

Figure 12.1 The concentrations of the principal carotenoids in human serum.

equally clear that an in-depth understanding of the human physiology of the oxy-carotenoids will be essential for the development of comprehensive answers as they apply to each specific carotenoid.

Important criteria which determine whether carotenoids individually or collectively are required by humans for normal, optimal health can be itemized Table 12.1. An essential dietary component must be universally present in all human populations. This is a surprisingly high hurdle to overcome and the conclusive data needed are not currently available for all human populations, much less for all carotenoids. The American dietary intake of carotenoids is not necessarily representative of that of other human populations. Therefore we cannot conclude that the carotenoids characteristically observed in human serum in the U.S. will be universally seen in all populations throughout the world. The seasonal availability of vegetable sources of carotenoids and the diversity of diets around the world add to the challenges of such investigations. Even if it is accepted that a dietary component is universally available in the human diet, there remains a considerable amount of physiological evidence required to establish unambiguously that it has an essential, functional role. The ability of humans to accumulate a carotenoid from the dietary sources at concentrations that are significant to a proposed function in a specific tissue, or cell types within a given tissue, must be established. Accumulation of a carotenoid in specific identifiable tissues, or cell types, by active transport is strongly supportive of the argument that the carotenoid serves a functional role, but passive nonspecific accumulation does not argue against the possibility of required function. That a given carotenoid is

TABLE 12.1
Evidence Needed to Establish a Functional Requirement for Carotenoids in Humans Must Satisfy Three Criteria

1. The carotenoid must be universally available in the diet of all human populations.
2. The carotenoid must be accumulated (either by active or passive transport) at concentrations consistent with the proposed function.
   A. Evidence of active transport
   B. Identification of regional sites of accumulation within specific tissues or subcellular organelles
3. Evidence must exist that a pathological condition is associated with a deficiency of the carotenoid.

essential ultimately must be established by the associated pathology or dysfunctional condition that occurs in its absence.

Whether or not any of the oxycarotenoids found in humans unambiguously meets these requirements remains to be established. As we briefly review the role of the four title carotenoids, we should look for not only evidence that they may play an essential, functional role, but also evidence that they may play a more vague nonspecific, yet nevertheless functional and advantageous, role in human physiology.

ASTAXANTHIN

Astaxanthin is an abundant oxycarotenoid (Structure 12.1). It is found in green algae and in the plumage of many birds including the flamingo. It is bound to the carotenoproteins present in the carapace of crustacea, and it is present in the flesh of salmon (Britton, 1983). Salmon is certainly a well known, but occasional, human source of astaxanthin. Astaxanthin is not found in common food plants and consequently is not a regular dietary carotenoid. Nor is it found in detectable levels in normal human serum. In many animals astaxanthin is produced through a biosynthetic

Astaxanthin
3,3'-dihydroxy-β,β-carotene-4,4'-dione

Structure 12.1

pathway from β-carotene, lutein, or zeaxanthin. There is no evidence that this pathway exists in humans (Scheidt, 1990).

Martin et al. and others have shown that astaxanthin is a highly effective antioxidant carotenoid *in vitro* (Martin et al., 1999; Terao et al., 1989; Woodall et al., 1997; Rengel et al., 2000; Miki, 1991). The antioxidant behavior of many carotenoids is characterized by a U-shaped graph (Figure 12.2). The rate of substrate oxidation decreases with increasing carotenoid concentration up to a concentration that is roughly 1 mM, but reverses and increases at higher concentration. This is typical of carotenoids such as β-carotene, which are designated mixed antioxidant/prooxidant, and the extent of the prooxidant upturn is dependent upon oxygen partial pressure. Astaxanthin is a pure antioxidant and lacks the right-hand arm of the U-curve at high carotenoid concentrations.

It has been suggested that interception of peroxy radicals and singlet oxygen by astaxanthin could be more effective than β-carotene, particularly in tissues where oxygen partial pressures are high. Astaxanthin has been regarded as having exceptional antioxidant properties. The relative antioxidant ability of several carotenoids is shown in Figure 12.3 (Martin et al., 1999). The high antioxidant ability of astaxanthin is ascribed to the presence of two keto groups at the 4 and 4' positions. These are conjugated to the polyene chain of the carotenoid through the 5,6 and 5',6' double bonds in the respective ionone rings. Several investigations using animal models, and a few investigations in humans, have been conducted to determine the ability of astaxanthin to function as an antioxidant.

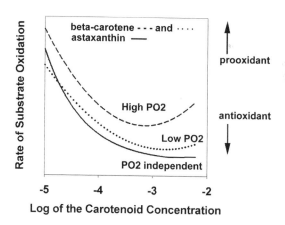

Figure 12.2 These qualitative curves illustrate the comparative antioxidant abilities of β-carotene and astaxanthin. The U-shape of the β-carotene curve (dashed) at high partial pressures of oxygen seen here is consistent with antioxidant function at low carotenoid concentrations. This function undergoes a transition at higher concentrations of oxygen to participate in chain-propagating oxidation steps characterized by an increased rate of substrate oxidation. At low partial pressures of oxygen, β-carotene (dotted) shows little prooxidant behavior. Astaxanthin (solid) is almost independent of the oxygen partial pressure. (Adapted from Martin, H.D. et al., *Pure Appl. Chem.*, 71, 2253–2262, 1999. With permission.)

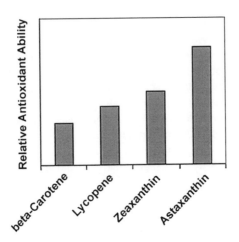

Figure 12.3 Relative antioxidant ability of β-carotene, lycopene, zeaxanthin, and astaxanthin illustrating the greater ability of carotenoids with oxygen functionalities to effectively disrupt the chain-propagation mechanism of peroxy radical oxidation. (Adapted from Martin, H.D. et al., *Pure Appl. Chem.*, 71, 2253–2262, 1999. With permission.)

HUMAN STUDIES

There is little published work documenting the role of astaxanthin in humans. Humans fed supplemental astaxanthin accumulate the carotenoid in the lipoproteins present in circulating blood. Serum levels of astaxanthin peaked at 1.2 μg/mL 6 hours after supplementation of subjects with a 100 mg dose of astaxanthin (Østerlie et al., 1999). 13-Z-astaxanthin was preferentially absorbed relative to all-E or the 9-Z isomers. The effect of geometrical isomerism on solubility may play a role in determining the extent of absorption in humans. The astaxanthin was equally distributed between the lipoprotein fractions (HDL, LDL, and VLDL) within the serum. (It should be noted that this is an extraordinarily large dose of carotenoid when compared to natural dietary intake for other carotenoids.) Astaxanthin has also been administered to *Heliobacter pylori*-positive nonulcer dyspeptic patients. A 21-day period of treatment of 10 patients with 8 mg of astaxanthin 5 times per day resulted in improvement of heartburn symptoms (8 of 10) and epigastric pain symptoms (7 of 10) at the end of an 8-week assessment period (Lignell et al., 1999).

ANIMAL STUDIES

Similarly, *H. pylori*-infected mice responded to astaxanthin treatment with a reduction in gastric inflammation (Bennedsen et al., 1999). Mice given astaxanthin also showed increased cytokine release from splenocytes. Nishino et al. have studied the ability of astaxanthin to inhibit spontaneous liver tumorigenesis in C3H/He mice (Nishino et al., 1999). Fifteen control mice were compared to an equal number of mice treated with astaxanthin (0.2 mg in 0.2 mL of corn oil, intragastric lavage, 3 times per week for 40 weeks). They found that treatment reduced the number of tumors from 0.87 in controls to 0.27 in treated animals ($p < 0.05$). The percentage of mice with tumors decreased from 53% in controls to 27% in treated animals. Chew has reported that astaxanthin was effective in limiting the growth of mammary tumors in mice (Chew et al., 1999). Tso and Lam treated rats with astaxanthin and found that they were less prone to photic damage than control animals, implying uptake of astaxanthin within the rat retina (Tso and Lam, 1996). Bendich reports that treatment of rats with astaxanthin produced an increase in HDL levels within the blood (Bendich, 1990).

These results are suggestive that the *in vivo* antioxidant function of astaxanthin may exert a positive influence on health status. Astaxanthin is clearly not an essential dietary component, nor does it have a normal functional role in humans, or for that matter, most animals. It nevertheless may be able to significantly reduce free radical damage in some tissues and its influence and effects should be further investigated. It is important to determine effects of different astaxanthin dosages, to what tissues astaxanthin is transported, and to what levels it accumulates in tissues.

β-CRYPTOXANTHIN

HUMAN STUDIES

β-Cryptoxanthin is an asymmetrically substituted hydroxycarotenoid (Structure 12.2). Tangerines, mangoes, and red bell peppers are some of the common human food sources rich in β-cryptoxanthin (Scott and Hart, 1994). Other more common food substances including corn also contain β-cryptoxanthin, although at lower levels. Based upon an estimated average human dietary intake of about 0.03 mg/day of β-cryptoxanthin for the American population, it appears that this carotenoid is more effectively absorbed than other major serum carotenoids. Normal serum levels of β-cryptoxanthin are ca. 0.15 μg/mL (Khachik et al., 1992). It would appear, at least in the American population, that β-cryptoxanthin is a universal component of the diet and consequently the serum as well. Humans are not known to accumulate

β-Cryptoxanthin
3-hydroxy-β,β-carotene

Structure 12.2

β-cryptoxanthin specifically in any particular tissue, and it is not identified with any known deficiency or physiologic disorder. On the basis of these data, it would be hard to argue that β-cryptoxanthin serves a uniquely essential function in humans. It is well known, however, that β-cryptoxanthin is a provitamin A carotenoid and is readily converted into 1 equivalent of vitamin A by cleavage between the 15 and 15' carbons (Bauernfeind et al., 1971). In addition, it has been identified as one of the xanthophylls present in human skin where it may be protective against sunburn (Stahl et al., 1999).

A very interesting and potentially significant report demonstrates that β-cryptoxanthin functions as an antioxidant. Incubation of cultured human lymphocytes with β-cryptoxanthin was followed by treatment with H_2O_2 and subsequently assayed for DNA damage using the Comet Assay (Astley et al., 1999). The Comet Assay detects single-strand breaks in the DNA resulting from peroxy radical attack. The conclusion of these investigators was that β-cryptoxanthin, as well as β-carotene and lutein, increases DNA resistance to H_2O_2 attack, and increases the repair rate for single-strand breaks induced by H_2O_2. In a similar study of 37 women by Haegele et al., serum β-cryptoxanthin concentration was inversely correlated with the level of 8-hydroxy-2'-deoxyguanosine (a product of DNA oxidation) detected in human lymphocytes isolated from serum and urine samples (Haegele et al., 2000).

There have been no human studies that document supplementation with purified β-cryptoxanthin, but several epidemiological studies show an association exists between serum or dietary levels of β-cryptoxanthin and disease conditions. In a study conducted by Franke et al., the association between a number of serum components, including carotenoids and thiobarituric acid-reactive substances (TBA-RS), was investigated (Franke et al., 1994). (TBA-RS are considered to be a measure useful in estimating lipid peroxide levels, which in turn are implicated as the cause of cellular oxidative damage including DNA cleavage and lipid peroxidation.) Their study indicates that there is a statistically significant inverse correlation between TBA-RS and several carotenoids, including β-cryptoxanthin (r = -0.38, p = 0.006), lutein/zeaxanthin (r = -0.61, p = 0.0001), and α- and β-carotene (r = -0.67, p = 0.0001).

β-Cryptoxanthin intake has been observed to be inversely associated to the incidence of lung cancer (Voorrips et al., 2000). In another study, carotenoid levels in lymphocytes from lung cancer patients were compared to those in healthy subjects (Schut et al., 1997). They observed a statistical association between reduced levels of lutein, lycopene, and α-carotene and lung cancer, but not between β-cryptoxanthin and lung cancer. In 528 elderly Dutch subjects studied, no association between lung function and serum β-cryptoxanthin was observed, though associations were observed with α- and β-carotene and lycopene (Grievink et al., 2000).

In a study of Crohn's disease, a chronic intestinal inflammatory condition, a group of 28 patients was compared to a control group of 23 healthy subjects. In the study, Crohn's disease was found to correlate with statistically lower levels of lutein, zeaxanthin, α-, and β-carotene, but not β-cryptoxanthin (Rumi et al., 2000). In cord blood samples obtained from 96 infants it was observed that β-carotene levels, but not β-cryptoxanthin or retinol, were lowered by maternal smoking (Moji et al., 1995). Among breast cancer patients (n = 266) studied in India, lowered levels of β-cryptoxanthin and other carotenoids were observed compared to levels in healthy subjects (Ito et al., 1999). For cirrhosis patients (n = 18), compared to controls (n = 20), a decrease of β-cryptoxanthin was observed in both serum and liver tissue for the former (p < 0.01) (Leo et al., 1993). (It was noted that cirrhosis patients having normal serum levels had depleted liver carotenoid levels, suggesting that cirrhosis interferes with liver uptake of carotenoids and normal carotenoid metabolism.)

In an investigation of the influence of alcohol on serum carotenoid levels, no statistical effect was detected between serum carotenoids (including β-cryptoxanthin) in alcholics (n = 95) before or after withdrawal from alcohol consumption or in comparison with control subjects (n = 118) (Lecompte et al., 1994). For women having high-risk type human papilloma virus, the adjusted mean concentration of serum carotenoids including β-cryptoxanthin was on average 24% lower for those in the highest risk categories (Guiliano et al., 1997). Among elderly patients with type II diabetes, serum levels of the carotenoids β-cryptoxanthin (p < 0.02), β-carotene (p < 0.01), and lycopene (p < 0.002) were inversely associated with age (Polidori et al., 2000).

ANIMAL STUDIES

The direct influence of β-cryptoxanthin intake on the development of colon tumors has been studied in F344 rats (Narisawa et al., 1999). N-methyl-nitrosourea-induced colon cancer in rats is significantly reduced for those given high doses of β-cryptoxanthin (25 ppm dietary level). Twenty-four (96%) of 25 control rats compared to only 17 (68%) of 25 β-cryptoxanthin-supplemented rats developed tumors during the 30-week study (p < 0.02). Lower doses of 1 or 5 ppm did not produce a statistically significant reduction in the number of tumors observed in similar study groups. Tissue levels were significantly elevated for the high-dose animals but not for those given the lower doses. Another rodent study (Nishino et al., 1999) investigated the effect of β-cryptoxanthin against tumorigenesis in mouse skin. Skin cancer was induced in 15 SENCAR mice, which were consuming β-cryptoxanthin at a level of 25 ppm in drinking water, by treatment with peroxynitrite and promotion with TPA (12-O-tetradecanoylphorbol-13-acetate). Comparison to 15 control mice revealed a modest, statistically significant reduction in the number of tumors among the β-cryptoxanthin-supplemented animals. A similar experiment with ICR mice in which skin cancer was induced with dimethyl benz[a]anthracene [DMBA] showed that β-cryptoxanthin administration resulted in a 29% tumor rate compared to a 64% rate in the mice not supplemented with the carotenoid. The authors concluded that β-cryptoxanthin is effective at inhibiting tumor formation and that it functions by stimulating the expression of the RB oncogene (Nishino et al., 2000).

Effective gap-junctional communication is an important functional characteristic in healthy multicellular tissues (Stahl et al., 1997; Stahl et al., 1998; Zhang et al., 1991). Gap junctions are direct channels connecting adjacent cells, which permit transfer between cells of molecules having masses below about 1000 amu, and are formed by a class of proteins referred to as connexin proteins. Gap-junctional communication is important in exchange of nutrients, ions, and signaling compounds.

There is a developing understanding that gap-junctional communication, or the lack thereof is important to the mechanisms of carcinogenesis and teratogenesis. A number of carotenoids are effective at stimulating gap-junctional communication, including β-cryptoxanthin. Importantly, the ability of carotenoids to induce gap-junctional communication is apparently unrelated to their antioxidant abilities.

It is apparent that there remains a considerable void in our knowledge of the influence of β-cryptoxanthin on human health and disease processes. Despite this, a fairly significant body of evidence supports the hypothesis that β-cryptoxanthin is effective as an antioxidant, and that it may play a role either directly or indirectly in the stimulation of important cell regulatory functions including cell gap-junctional communication and possibly oncogene stimulation.

LUTEIN AND ZEAXANTHIN

Lutein is the dominant xanthophyll in leafy green and yellow vegetables that are the primary human sources of carotenoids. Not surprisingly, lutein is also the dominant xanthophyll in human serum. Zeaxanthin is a constitutional isomer of lutein, and it differs from lutein structurally in subtle but important ways (Structures 12.3 and 12.4) (Landrum and Bone, 2001). Lutein has one β and one ε end-group whereas

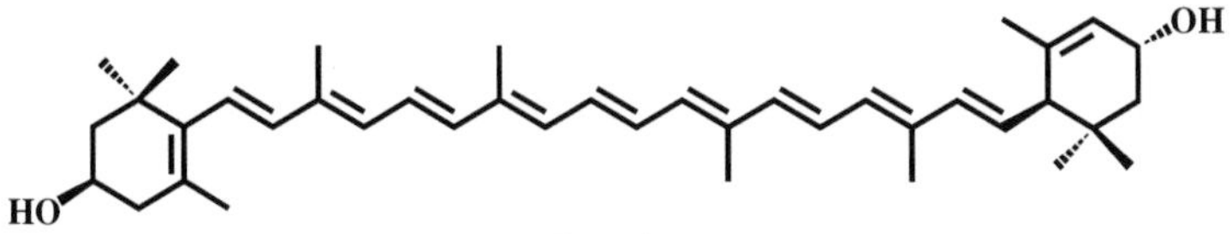

Lutein
3,3'-dihydroxy-β,ε-carotene

Structure 12.3

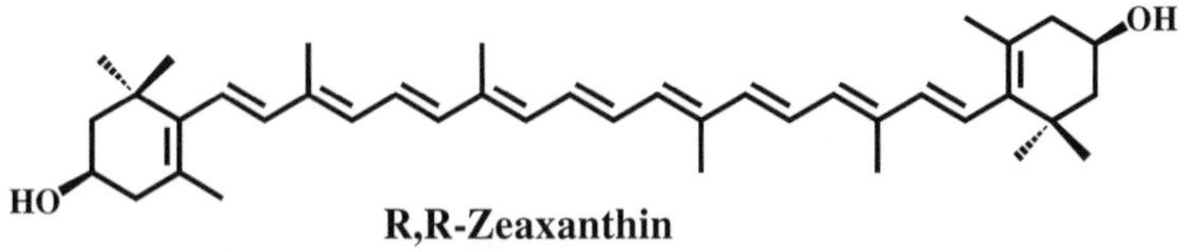

R,R-Zeaxanthin
(3R,3'R)-3,3'-dihydroxy-β,β-carotene

Structure 12.4

zeaxanthin is symmetric and has two β end-groups. Both lutein and zeaxanthin are dihydroxy carotenoids with the hydroxyl groups located on the 3 and 3' carbons. In lutein, the hydroxyl group is allylic to the isolated double bond in the ε-ring. Rich sources of lutein include spinach, kale, and broccoli. Zeaxanthin is the abundant xanthophyll in only a small number of food sources. In some of these, zeaxanthin is the dominant xanthophyll; they include orange peppers and Gou Zi Qi or lycium mill (*Lycium chinense*) berries, probably the richest source of zeaxanthin (Scott et al., 1994; Khachik et al., 1995).

The average combined lutein and zeaxanthin dietary intake is estimated to be 1.3–3 mg/d in the normal American diet (Nebling et al., 1997). The individual intake of lutein and zeaxanthin is dependent on dietary preferences, and the lutein to zeaxanthin ratio in the diet is not well established. Serum levels of lutein and zeaxanthin are often reported as a single combined value because separation of these isomers by HPLC is often impractical if the nonpolar hydrocarbons are simultaneously analyzed. Combined serum lutein and zeaxanthin concentration averages between 0.14 and 0.26 μg/mL, depending upon the study population (Yong et al., 1994; Khachik et al., 1992). The lutein to zeaxanthin ratio in human serum is somewhat variable and ranges from 2.5:1.0 to 4.0:1.0 (Landrum and Bone, 2000). Typical concentrations of serum lutein and zeaxanthin are 0.14 μg/mL and 0.05 μg/mL, respectively (Landrum et al., 1999).

HUMAN STUDIES

Lutein and zeaxanthin are uniquely accumulated within the macula of the human retina. The concentration of the macular pigment in the inner layers of the retina can be as great as 1 mM (Landrum et al., 1997). There is no site in the human body, with the possible exception of the corpus luteum, where exceptionally high carotenoid concentrations are observed. (Note: The bovine corpus luteum is well studied and the concentration of β-carotene is reported to be 2.3 μg/g in the corpus luteum itself, but is as great as 0.16 mg/g in the orange pigmented corpuscles of the ovary. There is no mention of significant levels of xanthophylls in the corpus luteum, nor have human studies been reported (Kirsche et al., 1987).) This is significant and is suggestive that the macular pigment must be accumulated via a mechanism that requires active transport and/or binding within the retina (Bone et al., 1993). Recently, Bernstein et al. have presented evidence that proteins within the retina bind lutein and zeaxanthin (Bernstein et al., 1997; Bernstein, 1999). Additional supportive evidence that the amount and composition of the macular pigment is actively controlled comes from the distribution of the two carotenoids within the retina (Bone et al., 1988).

In the central macula, zeaxanthin is the dominant carotenoid and exceeds the concentration of lutein by at least a factor of two; this ratio is inverted relative to the proportions found in the peripheral retina and also those in the diet and serum. Indeed, meso-zeaxanthin, a less common stereoisomer of zeaxanthin (Structure 12.5), which is not present in the human diet, is found in the central macula in quantities similar to that of the dietary R,R-zeaxanthin. In fact, the lutein and meso-zeaxanthin proportions appear to be inversely related. We think this relationship is consistent with a metabolic interconversion of lutein to meso-zeaxanthin.

Khachik has identified a minor carotenoid component of the macula, 3-hydroxy-ε,ε-caroten-3'-one, also called oxolutein (Structure 12.6). Oxolutein is produced by oxidation of the allylic hydroxyl of lutein (Herrero et al., 2000; Khachik et al., 1997; 1998). We have also observed this compound in our studies of human retinal extracts. We hypothesize this compound may be chemically reduced in an efficient process in the macula to generate meso-zeaxanthin, thereby accounting for several features of the carotenoid distribution within the retina. In summary, lutein and zeaxanthin are effectively accumulated within the retina in a manner that is highly indicative of active physiological control.

Epidemiological studies have implicated the macular pigment in the reduction of risk for the neovascular form of age-related macular degeneration (AMD) (Seddon et al., 1994). Dietary intake of lutein and zeaxanthin is associated with a 43% reduction in risk of advanced neovascular AMD for individuals in the highest quintile of intake. Similarly, a study of serum levels reveals the same association (EDCCSG, 1993). More recently, through a study of postmortem retinal samples, we have demonstrated that higher retinal levels of macular carotenoids are associated with a reduction in the odds ratio for the occurrence of AMD (Bone et al., 2001). Further, we have demonstrated through supplementation studies that macular pigment optical density is related to serum lutein and zeaxanthin levels. High doses of lutein produce macular pigment increases in the range of 20–40% (Landrum et al., 1997). Indeed, it is even possible to modulate the level of macu-

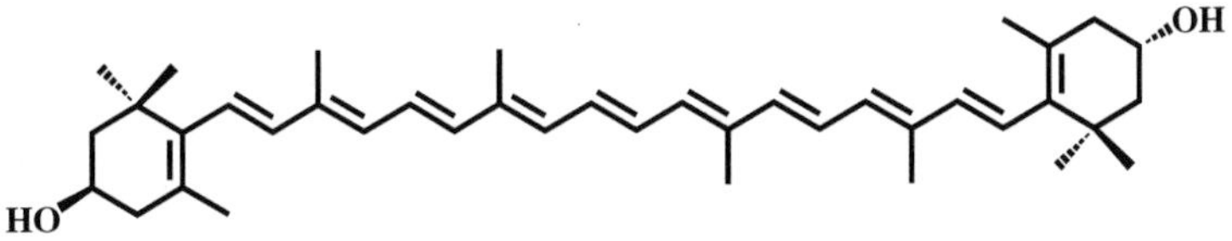

R,S-Meso-Zeaxanthin
(3R,3'S)-3,3'-dihydroxy-β,β-carotene

Structure 12.5

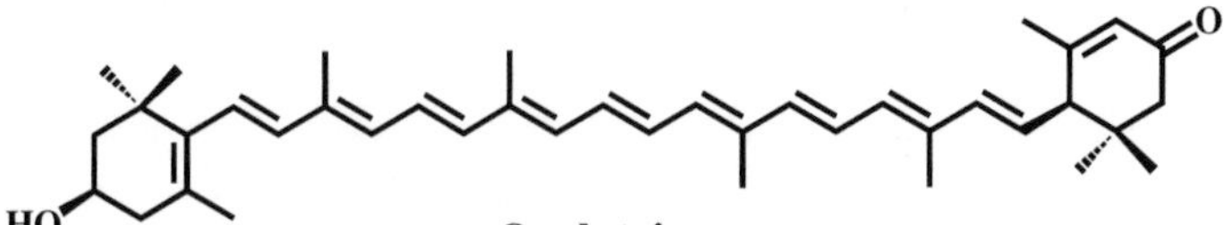

Oxolutein
3-hydroxy-β,ε-carotene-3'-one

Structure 12.6

lar pigment in human subjects who consume supplemental lutein (Landrum et al., 2000; Chen et al., 1999).

Lutein and zeaxanthin appear to be the only carotenoids that have been identified in human tissues that meet the basic criteria, described earlier, that might encourage us to characterize them as essential. They appear to be universally accumulated within the macula of humans and primates, where they may function by protecting against AMD. The mechanism by which the macular pigment functions to protect against AMD is probably twofold. Lutein and zeaxanthin in the inner retinal layers absorb in the range of 20–90% of the blue light incident on the retina, thereby reducing in a proportional extent the potential for photooxidative damage (Bone et al., 1992).

A second potential function for macular carotenoids involves their direct involvement in breaking the propagation steps involved in peroxy radical oxidation and/or interception of singlet oxygen that may be produced within the rods and cones themselves (Sommerburg et al., 1999; Rapp et al., 2000). Van Kuijk and Rapp have in separate investigations demonstrated the presence of lutein and zeaxanthin in rod outer-segment membranes, thereby placing these carotenoids in intimate contact with the structures most affected by the destructive effects of AMD. Lutein and zeaxanthin are effective antioxidants that have a lower prooxidant behavior than the hydrocarbon carotenoids, α- and β-carotene, at the high partial pressure of oxygen that is reported to be present in the central macula (Snodderly, 1995).

In addition to their highly specialized function in the macula, lutein and zeaxanthin may function as antioxidants, in what may be described as a generic manner, throughout the body. These carotenoids have, like β-cryptoxanthin, been demonstrated both by epidemiological studies and by animal and human studies to provide a protective function. Lutein and zeaxanthin protect against DNA damage, reducing both the number of single-strand breaks as demonstrated by the Comet Assay, and the amounts of 8-hydroxy-2'-deoxyguanosine formed in human lymphocytes exposed to oxidative stress (Haegele et al., 2000; Astley et al., 1999).

Epidemiological data collected for South Pacific Islanders found an association between a high intake of lutein and zeaxanthin and a lowering of the risk for lung cancer (Le Marchand et al., 1995). In the study conducted by Voorrips et al. 2000; (part of the Netherlands Cohort Study on Diet and Cancer, n = 58,279), intake of lutein and zeaxanthin (in addition to β-cryptoxanthin, folate, and vitamin C) was found to provide a protective effect against lung cancer. Lutein and zeaxanthin, like β-cryptoxanthin, are also both capable of stimulating gap-junctional communication (Stahl et al., 1998; Zhang et al., 1991). In a study of inflammatory response, lutein was shown to reduce inflammatory response in nonsmall-cell lung cancer patients as evidenced by an inverse correlation of C-reactive protein with serum lutein concentration ($p < 0.01$) (Talwar et al., 1997). In a case-control study (n = 231) of atherosclerosis, an inverse association between lutein and zeaxanthin levels and the extent of atherosclerosis was obtained (Iribarren et al., 1997).

ANIMAL STUDIES

In studies involving rodent tumorigenesis, Nishino et al. (1999) demonstrated that zeaxanthin can reduce spontaneous liver tumor formation in C3H/He mice when supplied at the level of 50 ppm in drinking water. Compared to controls (n = 14), of which 36% developed tumors, only 8% of zeaxanthin-treated mice (n = 12) were found to develop tumors. Rodent studies have also shown that lutein reduces the rate for occurrence of lung tumors promoted with 4-nitroquinoline-1-oxide in ddY mice (Nishino et al., 2000). Lutein-treated mice developed an average of 2.2 tumors per mouse compared to 3.1 tumors per mouse for untreated mice. Sprague-Dawley rats treated with the colon cancer initiator, N-methyl-nitrosourea, developed fewer aberrant crypt foci when treated with lutein as compared with control animals (Narisawa et al., 1999).

CONCLUSION

The evidence that the xanthophylls, particularly lutein and zeaxanthin but also β-cryptoxanthin, naturally function in part as generic lipophylic antioxidants in humans is a sound hypothesis based upon epidemiological data and animal studies. There is a compelling amount of evidence that lutein and zeaxanthin are essential components of the human macula, and that they function to reduce photooxidation in sensitive tissues. They may also be involved directly as antioxidants in the photoreceptors. Supplementation with lutein and zeaxanthin is a potentially practical therapeutic means to ensure that adequate levels of macular pigmentation are maintained to protect the retina and slow the processes which contribute to AMD.

Astaxanthin is a powerful antioxidant and may have great potential in reducing human disease processes that involve free radical mechanisms. Because there is a paucity of knowledge about astaxanthin metabolism in humans, extensive work must be done to determine what benefit can be realized by the incorporation of this carotenoid in the human diet. Indeed, care is warranted in the consumption of these carotenoids at levels that far exceed normal intake, until solid data establishing the desirability of such practices are available. This is especially true for smokers, as suggested by the Finnish ATBC study and the CARET study (Albanes et al., 1996; Redlich et al., 1999).

REFERENCES

Albanes, D., Heinonen, O.P., Taylor, P.R., Virtamo, J., Edwards, B.K., Rautalahti, M., Hartman, A.M., Palmgren, J., Freedman, L.S., Haapakoski, J.,Barrett, M.J., Pietinen, P., Malila, N., Tala, E., Liippo, K., Salomaa, E.R., Tangrea, J.A., Teppo, L., Askin, F.B., Taskinen, E., Erozan, Y., Greenwald, P., and Huttunen, J.K. Alpha-Tocopherol and beta-carotene supplements and lung cancer incidence in the alpha-tocopherol, beta-carotene cancer prevention study: effects of base-line characteristics and study compliance, *J. Natl. Cancer Inst.*, 88, 1560–1570, 1996.

Astley, S.B., Elliott, R.M., Archer, D., Hughes, D.A., and Southon, S. Carotenoids: DNA damage and repair balance, 12th Int. Carotenoid Symp., Cairns, Australia, abstract 6A-3, 1999.

Bauernfeind, J.C., Brubacher, G.B., Klaüi, H.M., and Marusich, W.L. Use of Carotenoids, in *Carotenoids* (Isler, O., ed.), Birkhauser Verlag, Basel, pp. 743–770, 1971.

Bendich, A. Carotenoids and the immune system, in *Carotenoid Chemistry and Biology* (Krinsky, N.I., ed.), Plenum Press, New York, 1990.

Bennedsen, M., Wang, X., Willén, R., Wadström, T., and Andersen, L.P. Treatment of *H. pylori*-infected mice with antioxidant astaxanthin reduces gastic inflammation, bacterial load and modulates cytokine release by splenocytes, *Immunol. Letts.*, 70, 185–189, 1999.

Bernstein, P.S. Purification of xanthophyll-binding proteins from the macula of the human retina, 12th Int. Carotenoid Symp., Cairns, Australia, abstract 3A-3, 1999.

Bernstein, P.S., Balashov, N.A., Tsong, E.D., and Rando, R.R. Retinal tubulin binds macular carotenoids, *Invest. Ophthalmol. Vis. Sci.*, 38, 167–175, 1997.

Bone, R.A., Landrum, J.T., Hime, G.W., Cains, A., and Zamor, J. Stereochemistry of the human macular carotenoids, *Invest. Ophthalmol. Vis. Sci.,* 14, 2033–2040, 1993.

Bone, R.A., Landrum, J.T., Mayne, S.T., Gomez, C.M., Tibor, S.E., and Twarawska, E.E. Macular pigment in donor eyes with and without AMD: a case-control study, *Invest. Ophthalmol. Vis. Sci.,* 42, 235-240, 2001.

Bone, R.A., Landrum, J.T., and Cains, A. Optical density spectra of the macular pigment *in vivo* and *in vitro, Vision Res.,* 32, 105–110, 1992.

Bone, R.A., Landrum, J.T., Fernandez, L., and Tarsis, S.L. Analysis of the macular pigment by HPLC: retinal distribution and age study, *Invest. Ophthalmol. Vis. Sci.,* 29, 843–849, 1988.

Britton, G. The Biochemistry of Natural Pigments, Cambridge University Press, Cambridge, 1983.

Britton, G., Liaaen-Jensen, S., and Pfander, H. Carotenoids Today and Challenges for the Future, in *Carotenoids* (Britton, G., Liaaen-Jensen, S., and Pfander, H., eds.), vol. 1, Birkhauser Verlag, Basel, pp. 13–26, 1995.

Chen, Y., Landrum, J.T., Bone, R.A., Dixon, Z., Bain, R., and Micah, S. Serum response to low-dosage supplementation with lutein, 12th Int. Carotenoid Symp., Cairns, Australia, 1999.

Chew, B.P., Park, J.S., Wong, M.W., and Wong, T.S. A comparison of the anticancer activities of dietary beta-carotene, canthaxanthin, and astaxanthin in mice *in vivo, Anticancer Res.,* 19, 1849–1853, 1999.

Demmig-Adams, B. and Adams, W.W. The role of xanthophyll cycle carotenoids in the protection of photosynthesis, *Trends in Plant Sci.*, 1, 21–26, 1996.

Eye Disease Case Control Study Group. Antioxidant status and neovascular age-related macular degeneration, *Arch. Ophthalmol.,* 111, 104–109, 1993.

Franke, A.A., Harwood, P.J., Shimamoto, T., Lumeng, S., Zhang, L.-X., Bertram, J.S., Wilkens, L.R., Le Marchand, L., and Cooney, R.V. Effects of micronutrients and antioxidants on lipid peroxidation in human plasma and in cell culture, *Cancer Lett.,* 79, 17–26, 1994.

Goodwin, T.W. Functions of Carotenoids, in *The Biochemistry of the Carotenoids*, Chapman and Hall, London, pp. 77–95, 1980.

Grievink, L., De Waart, F.G., Schouten, E.G., and Kok, F.J. Serum carotenoids, α-tocopherol, and lung function among Dutch elderly, *Am. J. Respir. Crit. Care Med.,* 161, 790–795, 2000.

Guiliano, A.R., Papenfuss, M., Mour, M., Canfield, L.M., Schneider, A., and Hatch, K. Antioxidant nutrients: associations with persistent human papilloma virus infection, *Cancer Epidemiol. Biomark. Prev.*, 6, 917–923, 1997.

Haegele, A.D., Gillette, C., O'Neill, C., Wolfe, P., Heimendinger, J., Sedlacek, S., and Thompson, H.J. Plasma xanthophyll carotenoid correlate inversely with indices of oxidative DNA damage and lipid peroxidation, *Cancer Epidemiol. Biomark. Prev.*, 9, 421–425, 2000.

Herrero, C., Chen, Y., Chi, J., Tibor, S., and Bone, R.A., and Landrum, J.T. Lutein metabolism in human serum and retina, *FASEB J.*, 14, A234, abstract 167.7, 2000.

Iribarren, C., Folsom, A.R., Jacobs, D.R., Jr., Gross, M.D., Belcher, J.D., and Eckfeldt, J.H. Association of serum vitamin levels, LDL susceptibility to oxidation, and autoantibodies against MDA-LDL with carotid atherosclerosis. A case-control study, *Arterioscler. Thromb. Vasc. Bio.*, 17, 1171–1177, 1997.

Ito, Y., Gajalakshmi, K.C., Sasaki, R., Suzuki, K., and Shanta, V. A study on serum carotenoid levels in breast cancer patients of Indian women in Chennai (Madras), India, *J. Epidemiol.*, 9, 306–314, 1999.

Khachik, F., Askin, F.B., and Lai, K. Distribution, bioavailability, and metabolism of carotenoids in humans, in *Phytochemicals: A New Paradygm* (Bidlack, W.R., Omaye, S.T., Meskin, M.S., Jahner, D., eds.), Technomic Publishing Company, Lancaster, Pennsylvania, 1998.

Khachik, F., Beecher, G.R., Goli, M.B., and Lusby, W.R. Separation, identification, and quantification of carotenoids in fruits, vegetables, and human plasma by high-performance liquid chromatography, *Pure Appl. Chem.*, 63, 71–80, 1991.

Khachik, F., Beecher, G.R., Goli, M.B., Lusby, W.R., and Smith, J.C., Jr. Separation, identification of carotenoids and their oxidation products in extracts of human plasma, *Anal. Chem.*, 64, 2111–2122, 1992.

Khachik, F., Beecher, G.R., and Smith, J.C., Jr. Lutein, lycopene, and their oxidative metabolites in chemoprevention of cancer, *J. Cell. Biochem.*, 22, 236–246, 1995.

Khachik, F., Bernstein, P.S., and Garland, D.L. Identification of lutein and zeaxanthin oxidation products in human and monkey retinas, *Invest. Ophthalmol. Vis. Sci.*, 38, 1802–1811, 1997.

Klaüi, H. and Bauernfeind, J.C. Carotenoids as Food Colors, in Carotenoids as Colorants and Vitamin A Precursors (Bauernfeind, J.C., ed.), Academic Press, New York, pp. 48–317, 1981.

Kirsche, B., Schlenzig, M., Ochrimenko, W.I., and Falchowsky, G. Influence of beta-carotene supplementation on carotene content of ovaries of heifers, *Arch. Tierernahr.*, 37, 995–999, 1987.

Krinsky, N.I. Function, in *Carotenoids* (Isler, O., ed.), Birkhauser Verlag, Basel, pp. 669–742, 1971.

Landrum, J.T. and Bone, R.A. Carotenoid nutrition and the human retina, *Int. J. Integ. Med.*, 2, 28–33, 2000.

Landrum, J.T. and Bone, R.A. Lutein, zeaxanthin and the macular pigment, *Arch. Biochem. Biophys.*, 385, 28-40, 2001.

Landrum, J.T., Bone, R.A., Chen, Y., Herrero, C., Llerena, C.M., and Twarowsk, E. Carotenoids in the human retina, *Pure Appl. Chem.*, 71, 2237–2244, 1999.

Landrum, J.T., Bone, R.A., Joa, H., Kilburn, M.D., Moore, L.L., and Spargue, K.E. A one year study of the macular pigment: The effect of 140 days of a lutein supplement, *Exp. Eye Res.*, 64, 57–62, 1997.

Landrum, J.T., Bone, R.A., and Kilburn, M.D. The macular pigment: a possible role in protection from age-related macular degeneration, *Adv. Pharmacol.,* 38, 537–556, 1997.

Le Marchand, L., Hankin, J.H., Bach, F., Kolonel, L.N., Wilkens, L.R., Stacewicz-Sapuntzakis, M., Bowen, P.E., Beecher, G.R., Laudon, F., Baque, P., Daniel, R., Seruvatu, L., and Henderson, B.E. An ecological study of diet and lung cancer in the South Pacific, *Int. J. Cancer,* 63, 18–23, 1995.

Lecomte, E., Grolier, P., Herbeth, B., Pirollet, P., Musse, N., Paille, F., Braesco, V., Siest, G., and Artur, Y. The relation of alcohol consumption to serum carotenoid and retinol levels. Effects of withdrawal, *Int. J. Vit. Nutr. Res.,* 64, 170–175, 1994.

Leo, M.A., Rosman, A.S., and Lieber, C.S. Differential depletion of carotenoids and tocopherol in liver disease, *Hepatol.,* 17, 977–986, 1993.

Lignell, A., Surace, R., Böttiger, P.A., and Borody, T.J. Symptom improvement in *Helicobacter pylori*-positive nonulcer dyspeptic patients after treatment with the carotenoid astaxanthin, 12th Int. Carotenoid Symp., Cairns, Australia, abstract 4A-3, 1999.

Martin, H.D., Ruck, C., Shcmidt, M., Sell, S., Beutner, S., Mayer, B., and Walsh, R. Chemistry of carotenoid oxidation and free radical reactions, *Pure Appl. Chem.,* 71, 2253–2262, 1999.

Miki, W. Biological functions and activities of animal carotenoids, *Pure Appl. Chem.,* 63, 141–146, 1991.

Moji, H., Murata, T., Morinobu, T., Mango, H., Tami, H., Okamot, R., Mino, M., Fujimura, M., and Takeuchi, T. Plasma levels of retinol, retinol-binding protein, all-trans beta-carotene and cryptoxanthin in low birth weight infants, *J. Nutr. Sci. Vitaminol.,* 41, 595–606, 1995.

Narisawa, T., Fukaura, Y., Oshima, S., Inakuma, T., Yano, M., Nishino, H. Chemoprevention by the oxygenated carotenoid β-cryptoxanthin of N-methyl-nitrosourea-induced colon carcinogenesis in F344 rats, *Jpn. J. Cancer Res.,* 90, 1061–1065, 1999.

Nebeling, L.C., Forman, M.R., Graubard, B.I., and Snyder, R.A. Changes in carotenoid intake in the United States: The 1987 and 1992 national health interview surveys, *J. Am. Diet. Assoc.,* 97, 991–996, 1997.

Nishino, H., Tokuda, H., Murakoshi, M., Stomi, Y., Matsumoto, H., Masuda, M., Bu, P., Onozuka, M., Yamaguchi, S., Okuda, Y., Takayasu, J., Nishino, A., Tsuruta, J., Okuda, M., Ichiishi, E., Nosaka, K., Konoshima, T., Kato, T., Nir, Z., Khackik, F., Misawa, N., Narisawa, T., and Taksuka, N. Cancer prevention by carotenoids and curcumin, in *Phytochemicals as Bioactive Agents* (Bidlack, W.R., Omaye, S.T., Meskin, M.S., and Topham, D.K.W., eds.), Technomic Publ. Co., Lancaster, Pennsylvania, pp.161–166, 2000.

Nishino, H., Tokuda, H., Satomi, Y., Masuda, M., Bu, P., Onozuk, M., Yamaguchi, S., Okuda, Y., Takayasu, J., Tsuruta, J., Okuda, M., Ichiishi, E., Murakoshi, M., Kato, T. Misawa, N., Narisawa, T., Takasuka, N., and Yano, M. Cancer prevention by carotenoids, *Pure Appl. Chem.,* 71, 2273–2278, 1999.

Østerlie, M., Bjerkeng, B., and Liaaen-Jensen, S. Blood appearance and distribution of E/Z isomers among plasma lipoproteins in humans administered a single meal with astaxanthin, 12th Int. Carotenoid Symp., Cairns, Australia, abstract 2A–13, 1999.

Polidori, M.C., Mecocci, P., Stahl, W., Parente, B., Cecchetti, R., Cherubini, A., Cao, P., Sies, H., and Senin, U. Plasma levels of lipophilic antioxidants in very old patients with type 2 diabetes, *Diabetes Metab. Res. Rev.,* 16, 15–19, 2000.

Rapp, L.M., Seema, S.S., and Choi, J.H. Lutein and zeaxanthin concentrations in rod outer segment membranes from perifoveal and peripheral human retina, *Invest. Ophthalmol. Vis. Sci.*, 41, 2100–2109, 2000.

Redlich, C.A., Chung, J.S., Cullen, M.R., Blaner, W.S., Van Bennekum, A.M., and Berglund, L. Effect of long-term beta-carotene and vitamin A on serum cholesterol and triglyceride levels among participants in the Carotene and Retinol Efficacy Trial (CARET), *Atherosclerosis*, 143, 427–434, 1999.

Rengal, D., Diez-Navajas, A., Serna-Rico, A., Veiga, P., Muga, A., and Milicua, J.C.G. Exogenously incorporated ketocarotenoids in large unilamellar vesicles. Protective activity against peroxidation, *Biochim. Biophys. Acta*, 1463, 179–187, 2000.

Rumi, G., Jr., Szabo, I., Vincze, A., Matus, Z., Toth, G., and Mozsik, G. Decrease of serum carotenoid in Crohn's disease, *J. Physiol. (Paris)*, 94, 159–161, 2000.

Schiedt, K. New Aspects of Carotenoid Metabolism in Animals, in *Carotenoids: Chemistry and Biology* (Krinsky, N.I., ed.), Plenum Press, New York, pp 247–268, 1990.

Schut, T.C.B., Puppels, G.J., Kraan, Y.M., Greve, J., van der Maas, L.I.J., and Figdor, C.G. Intracellular carotenoid levels measured by raman microscopy: Comparison of lymphocytes from lung cancer patients and healthy individuals, *Int. J. Cancer (Pred. Oncol.)*, 74, 20–25, 1997.

Scott, K.J. and Hart, D.J. The carotenoid composition of vegetables and fruits commonly consumed in the U.K., Norwich Laboratory, ISBN 0 7084 0549 5, 1994.

Seddon, J.M., Ajani, U.A., Sperduto, R.D., Hiller, R., Blair, N., Burton, T. C., Farber, M.D., Gragoudas, E.S., Haller, J., Miller, D.T., Yannuzzi, L.A., and Willett, W. Dietary carotenoids, vitamins A, C, and E and advanced age-related macular degeneration, *J. Am. Med. Assoc.*, 272, 1413–1420, 1994.

Snodderly, D.M. Evidence for protection against age-related macular degeneration by carotenoids and antioxidant vitamins, *Am. J. Clin. Nutr. 26* (suppl.), 1448S–1461S, 1995.

Sommerburg, O.G., Siems, W.G., Hurst, J.S., Lewis, J.W., Kliger, D.S., and van Kuijk, F.J.G. M. Lutein and zeaxanthin are associated with photoreceptors in the human retina, *Curr. Eye Res.*, 19, 491–495, 1999.

Stahl, W. and Sies, H. The role of carotenoids and retinoids in gap-junctional communication, *Int. J. Vit. Nutr. Res.*, 68, 354–359, 1998.

Stahl, W., Heinrich, U., Jungmann, H., Sies, H., and Tronnier, H. Carotenoids in human skin — protective effects against sunburn, 12th Int. Carotenoid Symp., Cairns, Australia, abstract 4A-1, 1999.

Stahl, W., Nicolai, S., Briviba, K., Hanusch, M., Broszeit, G., Peters, M., Martin, H.-D., and Sies, H. Biological activities of natural and synthetic carotenoids: induction of gap-junctional communication and singlet oxygen quenching, *Carcinogenesis*, 18, 89–92, 1997.

Straub, O. *Key to Carotenoids, 2nd ed.* (Pfander, H., ed.), Birkhauser Verlag, Basel, 1987.

Talwar, D., Ha, T.K., Scott, H.R., Cooney, J., Fell, G.S., O'Reilly, D.S., Lean, M.E., and McMillan, D.C. Effect of inflammation measures of antioxidant status in patients with nonsmall cell lung cancer, *Am. J. Clin.Nutr.*, 66, 1283–1285, 1997.

Terao, J. Antioxidant activity of β-carotene-related carotenoids in solution, *Lipids*, 24, 659–661, 1989.

Tso, M.O.M. and Lam, T.T. Method of retarding and ameliorating central nervous system and eye damage, U.S. Patent 5,527,533, 1996.

Voorrips, L.E., Goldbohn, R.A., Brants, H.A.M., van Poppel, G.A.F.C., Sturmans, F., Hermus, R.J.J., and van den Brandt, P.A. A prospective cohort study on antioxidant and folate intake and male lung cancer risk, *Cancer Epidemiol. Biomark. Prev.*, 9, 357–365, 2000.

Weedon, B.C.L., Occurrence, in *Carotenoids* (Isler, O., ed.), Birkhauser Verlag, Basel, pp. 29–59, 1971.

Woodall, A.A., Lee, S.W.-M., Weesie, R.J., Jackson, M.J., and Britton, G. Oxidation of carotenoids by free radicals: relationship between structure and reactivity, *Biochem. Biophys. Acta,* 1336, 33–42, 1997.

Yong, L.-C., Forman, M.R., Beecher, G.R., Graubard, B.I., Campbell, W.S., Reichman, M.E., Taylor, P.R., Lanza, E., Holden, J.M., and Judd, J.T. Relationship between dietary intake and plasma concentrations of carotenoids in premenopausal women: application of the USDA-NCI carotenoid food-composition database, *J. Clin. Nutr.,* 60, 223–230, 1994.

Zhang, L.X., Cooney, R.V., and Bertram, J.S. Carotenoids enhance gap junctional communication and inhibit lipid peroxidation in C3H/10T1/2 cell: relationship to their cancer chemopreventive action, *Carcinogenesis,* 12, 2109–2114, 1991.

Index